T0271357

# Pests and Diseases in Vegetable Crops

**Dr. S. Parthasarathy**
Assistant Professor (Plant Pathology)
Amrita School of Agricultural Sciences
Amrita Vishwa Vidyapeetham, Coimbatore

**Dr. P. Lakshmidevi**
Teaching Assistant (Plant Pathology)
Institute of Agriculture, Kumulur, Trichy
Tamil Nadu Agricultural University, Coimbatore

**Dr. P. Yashodha**
Associate Professor (Entomology)
Agricultural College and Research Institute, Thiruvannamalai
Tamil Nadu Agricultural University, Coimbatore

**Dr. C. Gopalakrishnan**
Professor (Plant Pathology)
Department of Plant Pathology
Centre for Plant Protection Studies
Tamil Nadu Agricultural University, Coimbatore

CRC Press
Taylor & Francis Group
Boca Raton London New York

CRC Press is an imprint of the
Taylor & Francis Group, an **informa** business

–EPH–
**Elite Publishing House**

First published 2025
by CRC Press
4 Park Square, Milton Park, Abingdon, Oxon, OX14 4RN

and by CRC Press
2385 NW Executive Center Drive, Suite 320, Boca Raton FL 33431

*CRC Press is an imprint of Informa UK Limited*

© 2025 Elite Publishing House

The right of S. Parthasarathy, P. Lakshmidevi, P. Yashodha and C. Gopalakrishnan to be identified as authors of this work has been asserted in accordance with sections 77 and 78 of the Copyright, Designs and Patents Act 1988.

All rights reserved. No part of this book may be reprinted or reproduced or utilised in any form or by any electronic, mechanical, or other means, now known or hereafter invented, including photocopying and recording, or in any information storage or retrieval system, without permission in writing from the publishers.

For permission to photocopy or use material electronically from this work, access www.copyright. com or contact the Copyright Clearance Center, Inc. (CCC), 222 Rosewood Drive, Danvers, MA 01923, 978-750-8400. For works that are not available on CCC please contact mpkbookspermissions@tandf.co.uk

*Trademark notice*: Product or corporate names may be trademarks or registered trademarks, and are used only for identification and explanation without intent to infringe.

Print edition not for sale in South Asia (India, Sri Lanka, Nepal, Bangladesh, Pakistan or Bhutan).

*British Library Cataloguing-in-Publication Data*
A catalogue record for this book is available from the British Library

ISBN: 9781032823683 (hbk)
ISBN: 9781032823706 (pbk)
ISBN: 9781003504153 (ebk)

DOI: 10.4324/9781003504153

Typeset in Adobe Caslon Pro
by Elite Publishing House, Delhi

-EPH-

# Contents

# Preface

Management of pests and diseases are greatest importance for successful and profitable cultivation of crops. Under these environments, important and relevant information were collected and compiled in a textbook form titled Pests and Diseases in Vegetable Crops. This book is mainly intended for the Plant Protection courses of graduate students in the field of Agriculture, Horticulture, Botany, Forestry and Zoology. It is clear that young students are suffering from cultural shocks to shift from their environment. Semester system of education of B.Sc./M.Sc. (Ag.), B.Sc./M.Sc. (Horti.), B.Sc./M.Sc. (Forestry), B.Sc./M.Sc. (Botany), and B. Tech. (Horti.), students are quite dynamic for which the students are to be helped for changeover. We can identify their difficulties for comprehension of language, non-availability of textbooks for their semester system. There is a need to use simple language. The present book titled Diagnostics and Management of Pest and Diseases in Vegetable Crops suite to the need of students. This book is written in simple understandable language dealing with various subject matters of Pests and Diseases. This book has been prepared with a specific purpose of importing complete comprehensive information about major diseases of horticultural crops in India and we hope that the students and readers will find this with much utility. This book aims to project the significant pest, disease, and nematode issues affecting essential vegetable crops in the country. The management of pests in vegetable crops has been extensively addressed through many methods, including regulatory, physical, cultural, chemical, and biological approaches, host resistance, and integrated pest management strategies.

This book will function as a comprehensive manual for individuals involved in identifying and mitigating pest-related issues in vegetable crops inside agricultural fields. Additionally, it can function as a pragmatic manual for farmers who cultivate vegetable crops. Moreover, this resource is a valuable reference point for policymakers, researchers, extension workers, and students. The book possesses potential utility in instructing both undergraduate and postgraduate curricula about crop protection.

I thank all the scientists / publishers from which references were collected on various aspects of pests and disease aspects. I am sure that this book will serve as valuable text cum reader friendly textbook to the graduate and post-graduate students of agricultural universities.

**Authors**

# About the Authors

**Dr. S. Parthasarathy** is working as an Assistant Professor (Plant Pathology), in the Department of Plant Pathology at Amrita School of Agricultural Sciences. He has completed his PG and Ph.D. from Tamil Nadu Agricultural University, Coimbatore. He is a recipient of innovative fellowships such as the BIRAC Post Master's Innovation Fellow in DBT during 2015–17, the Canadian International Development Research Centre-Nano Project JRF during 2013–14, the UGC Non-Special Assistance Programme during 2012, and the Ministry of Human Resource Development Fellowship for UG during 2009, and he also received medals during his UG program and several meritorious awards in international and national conferences for his research contributions. A budding expert in Plant Pathology, he has published several research articles in refereed journals, authored 16 books, and contributed about 45 chapters to different books with national and international reputations. He also served as a member of the editorial boards of national journals and several international scientific societies. His areas of specialization are Biological Control and Molecular Plant Pathology.

**Dr. P. Lakshmidevi** is a Teaching Assistant (Plant Pathology) in the Department of Plant Pathology, Institute of Agriculture. She has completed her Ph.D. from Tamil Nadu Agricultural University, Coimbatore. She is a recipient of innovative fellowships, a Fellow in RGNF from 2011-14, a Young Plant Pathologist 2019, and the Best Book Contributor Award 2022. She has conducted 2 Farm Testing trials, 4 Front Line Demonstrations, two farmers' field schools, and 43 on and off-campus training. She has published 4 books with ISBNs, 12 research articles, 29 conference papers, and 23 book chapters with ISBNs.

**Dr. P. Yashodha** is an Associate Professor (Entomology), at the Department of Entomology, Agricultural College and Research Institute, Thiruvannamalai. She has completed 14 years of service and handled numerous undergraduate and postgraduate courses with a specialization in Biological controls, Insecticide Toxicology, Storage Entomology, and Molecular Ecology. She has guided many M.Sc. and over 5 Ph.D. students in Tamil Nadu Agricultural University. Numerous academic organizations have awarded her for her outstanding contributions to Agricultural Entomology. She has contributed to releasing crop varieties and technologies in Tamil Nadu. She has operated numerous funded projects in the university, including the DST and NAIP projects. She has published over 3 books with ISBNs, 40 research articles, 25 conference papers, 18 book chapters with ISBNs.

**Dr. C. Gopalakrishnan** is working as a Professor of Plant Pathology at Tamil Nadu Agricultural University, Coimbatore. He has completed 28 years of service and has handled more than 50 courses at the undergraduate level and 34 courses at the postgraduate level. He has guided 3 Ph.D. and 7 PG students. Five Ph.D. students are currently working under his guidance in various areas of Plant Pathology. He has authored 6 books, 13 book chapters and published more than 50 research papers in national and international journals. He received a NATP-Team of Excellence fellowship in biological control during 2002-03. He has received one state-level and four university-level awards for his contributions to agricultural research, extension, and teaching. He has attended over 30 national and international conferences, seminars and symposia. He was involved in releasing 8 crop varieties and 3 technologies to benefit the farming community.

# 1 Introduction

The cultivation of vegetable crops is facing a significant challenge due to several destructive pests, including insect and mite pests, fungal, bacterial, viral/mycoplasma, and nematode diseases. The significant damages inflicted by these adversaries of agriculturalists have garnered the interest of professionals in the field of agriculture and decision-makers. In recent years, there has been a noticeable increase in the occurrence and severity of pest issues affecting vegetable crops. These problems can be attributed to changes in cropping patterns, the adoption of intensive cultivation practices, and a shift toward water-saving strategies. The inadequate presence of crop protection specialists within extension programs, both in the public and commercial sectors, has significantly influenced the accurate identification and diagnosis of pest-related issues. Accurately identifying pests affecting vegetable crops remains a significant challenge in the present era. The erroneous advice for pest management can be attributed to the incorrect identification of pests.

Diagnosing and managing pest and diseases in vegetable crops is crucial for ensuring a healthy and productive harvest. Diagnosing pests and diseases in vegetable crops is essential for implementing timely and effective management strategies. Here's a step-by-step guide to diagnostics:

1. **Regular Field Inspections:**
   » Conduct frequent and systematic field inspections. Walk through the vegetable fields and observe plants closely, paying attention to leaves, stems, fruits, and the overall plant health.

2. **Symptom Recognition:**
   » Learn to recognize common symptoms of pests and diseases. Symptoms may include wilting, yellowing, spots, holes, chewed leaves, distorted growth, or the presence of insects on plants.

3. **Sample Collection:**

   » If you spot unusual symptoms or suspect a particular pest or disease, collect samples for further examination. Take samples that represent both affected and healthy plant material.

4. **Laboratory Analysis:**

   » If available, consult with agricultural extension offices or diagnostic laboratories for accurate identification. They may analyze the samples to identify the specific pest, pathogen, or disease causing the issue.

5. **Online Resources and Field Guides:**

   » Utilize online resources, plant disease databases, and field guides to help identify common pests and diseases. Many agricultural universities and organizations offer online tools for identifying plant problems.

6. **Comparison with Reference Images:**

   » Compare the symptoms of your crops with reference images of known pests and diseases. This can aid in narrowing down the possible causes.

7. **Environmental Assessment:**

   » Consider environmental factors that could be influencing the plant health. Issues like poor drainage, overwatering, nutrient deficiencies, or extreme temperatures can sometimes mimic pest and disease symptoms.

8. **Insect Traps and Monitoring:**

   » Use insect traps and monitoring tools to detect and identify insect pests present in your vegetable fields.

9. **Professional Consultation:**

   » Seek advice from local agricultural experts, agronomists, or experienced farmers who have dealt with similar pest and disease issues in your region.

10. **Keep Records:**

   » Maintain detailed records of your field inspections, observations, and any treatments applied. This documentation will help you track the development of pests and diseases over time.

11. **Rapid Response:**

   » If a pest or disease is confirmed, take immediate action to manage it. Early intervention can prevent further spread and minimize crop damage.

12. **Implement Integrated Pest Management (IPM):**

» Employ an integrated approach to manage pests and diseases. Combine cultural practices, biological controls, and judicious use of pesticides to effectively control the problem while minimizing harm to beneficial organisms and the environment.

Remember that accurate diagnostics are the foundation of effective pest and disease management in vegetable crops.

Morphological and molecular diagnostics are essential tools for identifying and characterizing pests and diseases that affect plants. These diagnostic approaches help researchers, plant pathologists, and farmers understand the nature of the pest or disease and implement appropriate management strategies.

**1. Morphological Diagnostics:** Morphological diagnostics involve the visual examination of the external characteristics and physical structures of pests or disease-causing organisms. Here are the key steps involved in morphological diagnostics:

» Pest Identification: For pests, such as insects and mites, the process involves observing their body shape, size, color, antennae, legs, wings, mouthparts, and other distinctive features.

» Disease Symptom Observation: For diseases, morphological diagnostics entail observing the visible symptoms on the plants, such as leaf spots, wilting, necrosis, galls, cankers, discolorations, and other abnormal growth patterns.

» Sign Observation: Morphological diagnostics also involve identifying physical evidence of pests or pathogens, known as signs. Examples include insect eggs, larval stages, fungal spores, bacterial ooze, or nematode cysts.

» Sample Collection: Collect samples of plants or pests representing both healthy and affected parts. Ensure that the samples are from active infestation or disease areas.

» Microscopic Examination: Examine the collected samples under a microscope to observe the morphology and characteristics of pests, pathogens, or their structures.

» Comparative Analysis: Compare the observed characteristics with known features of different pests and pathogens to make a preliminary identification.

**2. Molecular Diagnostics:** Molecular diagnostics involve the use of molecular techniques to identify pests and pathogens at the genetic level. Here are the key steps involved in molecular diagnostics:

» Nucleic Acid Extraction: Extract DNA or RNA from the collected samples of pests or affected plant tissues. This step isolates the genetic material of the pest or pathogen.

» Polymerase Chain Reaction (PCR): Use PCR to amplify specific regions of the pest or pathogen's DNA or RNA. PCR amplifies the genetic material, making it easier to detect and identify.

» Gel Electrophoresis: Separate the amplified DNA fragments using gel electrophoresis. This step helps confirm the presence and size of the amplified DNA fragments.

» Sequencing: Sequence the amplified DNA fragments to compare them with known genetic sequences of pests and pathogens available in databases. This enables precise identification.

» Real-Time PCR (qPCR): qPCR allows for the quantitative measurement of the pest or pathogen's genetic material. It helps determine the severity of the infestation or disease.

» Next-Generation Sequencing (NGS): NGS techniques provide high-throughput sequencing, enabling the identification of multiple pests or pathogens in a single sample.

**Benefits of Morphological and Molecular Diagnostics:**

» Morphological diagnostics are relatively fast and cost-effective, making them suitable for initial identification in the field.

» Molecular diagnostics offer high specificity and sensitivity, allowing for accurate identification, especially in cases where pests or pathogens are difficult to distinguish based on morphology alone.

By combining both morphological and molecular diagnostic approaches, researchers and agricultural professionals can accurately identify pests and diseases, understand their characteristics, and implement targeted management strategies to protect crops effectively. By identifying the specific issues affecting your crops, you can implement appropriate and targeted solutions to protect your plants and ensure a successful harvest.

The management of pests and diseases in vegetable crops requires an integrated approach that combines various strategies to minimize the impact on crops while reducing reliance on chemical pesticides. Here are some general steps and guidelines to help you with the process.

1. **Regular Monitoring:**

» Regularly inspect your vegetable crops for any signs of pests or diseases. Walk through your fields and observe the plants closely, paying attention to leaves, stems, fruits, and the soil.

2. **Identification:**

   » Identify the specific pests and diseases affecting your vegetable crops. This can be done through visual inspection or with the help of agricultural experts, local extension offices, or diagnostic laboratories.

3. **Establishing Economic Thresholds:**

   » Determine the economic threshold level for each pest species. The economic threshold is the point at which the pest population reaches a level where control measures are economically justified.

4. **Integrated Pest Management (IPM):**

   » Adopt an integrated approach to manage pests and diseases. IPM involves combining various strategies to minimize the impact of pests while reducing the use of chemical pesticides. It includes cultural, biological, and chemical control methods.

5. **Mechanical Controls:**

   » Use physical barriers, traps, or handpicking to manage pests manually. For instance, you can use row covers to protect plants from certain insects or physically remove pests by hand.

6. **Pheromones and Traps:**

   » Implement pheromone traps to monitor and control certain pests. Pheromones can attract pests to traps, reducing their population.

7. **Cultural Controls:**

   » Implement a crop rotation plan to disrupt pest and disease life cycles. Avoid planting the same vegetable or closely related crops in the same area for consecutive growing seasons.

   » Practice good sanitation by removing and disposing of plant debris, weeds, and other potential hosts for pests and diseases. This helps reduce the source of infections and pest infestations.

   » Select vegetable varieties that have natural resistance to common pests and diseases prevalent in your region. Check seed catalogs or consult with local agricultural experts for suitable resistant varieties.

   » Maintain proper spacing between plants to improve air circulation and reduce disease spread.

   » Ensure adequate irrigation and drainage to avoid water-related issues that can lead to disease development.

» Use companion planting to create a diverse and balanced ecosystem that naturally deters pests. Certain plants can repel or attract specific insects, helping to protect your vegetable crops.

8. **Biological Controls:**

» Introduce beneficial insects and organisms that prey on or parasitize pests. Ladybugs, lacewings, and certain nematodes are examples of beneficial organisms used in biological control.

9. **Chemical Controls:**

» Use chemical pesticides only when necessary and as a last resort. Always follow the recommended dosage and application instructions to minimize negative impacts on the environment and non-target organisms.

» Rotate between different chemical groups to prevent the development of pesticide resistance.

10. **Organic and Natural Remedies:**

» Explore organic and natural remedies such as neem oil, garlic spray, or soap solutions to control certain pests and diseases.

11. **Record Keeping:**

» Keep detailed records of pest and disease occurrences, treatments applied, and their effectiveness. This information will help you make informed decisions in the future.

12. **Consult Experts:**

» Seek advice from local agricultural extension officers, experienced farmers, or agronomists to get tailored recommendations for your specific crop and region.

13. **Training and Education:**

» Stay updated with the latest research and developments in pest and disease management through workshops, seminars, and agricultural publications.

Remember that effective pest and disease management in vegetable crops requires continuous attention and proactive measures. By following these steps and continuously monitoring your vegetable crops, you can effectively diagnose and manage pests and diseases, helping to ensure a healthy and productive harvest. Early detection and intervention can prevent severe damage and ensure a successful vegetable harvest.

# References

Martin, R. R., James, D., & Lévesque, C. A. (2000). Impacts of molecular diagnostic technologies on plant disease management. Annual review of phytopathology, 38(1), 207-239.

Miller, S. A., Beed, F. D., & Harmon, C. L. (2009). Plant disease diagnostic capabilities and networks. Annual review of phytopathology, 47, 15-38.

De Waele, D., & Elsen, A. (2007). Challenges in tropical plant nematology. Annu. Rev. Phytopathol., 45, 457-485.

Gariepy, T. D., Kuhlmann, U., Gillott, C., & Erlandson, M. (2007). Parasitoids, predators and PCR: the use of diagnostic molecular markers in biological control of Arthropods. Journal of applied entomology, 131(4), 225-240.

Armstrong, K. F., & Ball, S. L. (2005). DNA barcodes for biosecurity: invasive species identification. Philosophical Transactions of the Royal Society B: Biological Sciences, 360(1462), 1813-1823.

Ciancio, A., & Mukerji, K. G. (Eds.). (2007). General concepts in integrated pest and disease management (No. 04; SB611. 5, G4.). New York, NY, USA:: Springer.

Gullino, M. L., Albajes, R., & Nicot, P. C. (Eds.). (2020). Integrated pest and disease management in greenhouse crops (Vol. 9). New York, NY, USA:: Springer International Publishing.

Gajanana, T. M., Moorthy, P. N., Anupama, H. L., Raghunatha, R., & Kumar, G. T. (2006). Integrated pest and disease management in tomato: an economic analysis. Agricultural economics research review, 19(2), 269-280.

# 2

# Amaranth

## INSECT PESTS

### Aphids, *Lipaphis erysimi*

#### Damage symptoms

- » Curled leaves brought on by aphids look bad to potential buyers.
- » Plant development is hindered because aphids eat on the plant's juices.
- » Leaf wrinkles, poor development, and misshapen seeds are common symptoms of heavily infested plants.
- » Plants, especially young plants, can dry out and die from a severe aphid infestation.
- » Damage to elder plants can reduce bloom and seed output, which can lead to a failed harvest.
- » Seed viability may also be impacted by damage.

#### Management

- » Maintain constant vigilance over the harvest.
- » Always only spray the affected plants unless otherwise instructed (spot spraying).
- » Using a spray of 4% NSKE is a good method for preventing aphid infestations.
- » Preserve the ecological balance by preserving pests. They play a crucial role in the prevention of aphid infestations by natural means.

**Leaf miner,** *Liriomyza huidobrensis*

**Damage symptoms**

» The maggot creates thin, white tunnels in the leaves.

» Mined leaves may turn yellow and fall if the damage is severe enough.

» When seedlings are severely damaged, they become stunted and may even perish.

» When attacks are severe, especially on young plants, control measures must be implemented.

**Pest identification**

» The fly species known as leaf miners measure only 1.3-1.6 mm in length.

**Management**

» Preserve the ecological balance by preserving pests. As such, they play a crucial role in the ecological management of leaf miners.

» Leaf mines should be destroyed by hand.

» Apply Neem pesticides to the crop as needed. Controlling leaf miners effectively with Neem oil or Neem water extracts.

**Stem weevil,** *Hypolixus haerens*

**Damage symptoms**

» Adult weevils eat leaves, but it›s the larvae (grubs) that do the real damage by boring into the plant›s roots and stems and setting the plant up for rot, lodging, and disease.

» The most devastating are the stem-boring weevils that cause plants to wilt and fall over.

» Stems are hollowed out by the larvae as they make their way to the root collar. Stems that have been fed on by larvae are more brittle in wind, which increases crop losses.

» Discoloration, rotting, and cankers in stems, roots, and branches have all been linked to weevil infestation.

» Disease-causing fungus (mostly *Fusarium* spp.), which this weevil spreads and causes tissue deterioration and canker, have been linked to this pest.

**Pest identification**

  » Eggs have a uniformly smooth exterior and an oval, light golden color.

  » A stout, curled, legless, white monster.

  » The adult is a dark grey with brown elytra and a fairly long snout.

**Management**

  » Round up the wild amaranth hosts in the area and wipe them out.

  » To lessen the presence of weevils and protect unaffected plants, uproot and destroy any infested plants.

  » When the leaves and stems have been harvested, spray the area with Malathion 50 EC 500 ml, Endosulfan 35 EC 500 ml, or Dichlorvos 375 ml in 500 litres of water per hectare. Prepare for the next harvest in 15–20 days.

**Leaf caterpillar,** *Hymenia recurvalis*

**Damage symptoms**

  » Leaf green caterpillars that munch on foliage.

  » Caterpillars make leaves unfit for food by weaving them together with silky threads, then trapping themselves within, whereupon they scrape the green section.

  » This causes the leaves' webs to dry out and fall off.

**Pest identification**

  » Snowy white spheres, eggs can be placed alone or in clutches of two to five in the troughs between leaf veins.

  » Larvae are olive-brown, with white lines and black crescents below the lateral lines of the thorax.

  » Adults of this species look like little black moths with long, lean bodies. Waving white lines contrast with the dark brown of the wings.

**Spread**

  » *H. recurvalis* moths are capable fliers.

  » Caterpillars feed as they go along the plant life.

## Management

» Round up the wild amaranth hosts in the area and wipe them out.

» Get rid of the grubs and adults, and then throw away the infected plant components.

» Light traps with a density of 1-2/ha can be used to lure in and kill the adults.

» When the leaves and stems have been harvested, spray the area with Malathion 50 EC 500 ml, Endosulfan 35 EC 500 ml, or Dichlorvos 375 ml in 500 L of water per ha. Prepare for the next harvest in 15–20 days.

» The leaf caterpillar was well managed by applying 'Thuricide' (*Bacillus thuringiensis*) dust ($3x10^6$ spores/g) at 20–25 kg/ha, 'Thuricide' WP (3x106 spores/g) at 0.1 and 0.2%, and 'Thuricide' EC ($15x10^9$ spores/g) at 0.1 and 0.2%.

» The parasitoid *Apanteles* sp. was discovered to have parasitized 62% of the leaf caterpillar population.

## Cutworms, *Agrotis* sp.

### Damage symptoms

» Little plants are easy targets for cutworms.

» During night, the caterpillar emerges from the ground, wraps its body around a plant, and snips its way through the stem of a tender young plant.

» As a result, subsurface vegetation could be harmed.

» Weakened by cutworm attack, plants quickly wither and perish.

» In most cases, the damage caused by cutworms is not severe enough to necessitate treatment.

» Nonetheless, in extreme cases, a new crop may be lost.

### Pest identification

» An adult caterpillar will measure between 3.5 and 5 cms in length.

### Management

» Keep an eye on the situation by tallying the number of young plants that have been injured or trimmed. The cutworm should be watched first thing in the morning.

» Get rid of and kill all cutworms.

» Make sure the land is weed-free and ready to receive the crop 10-14 days before planting.

» Caterpillars are vulnerable to predation and dehydration when ploughing disturbs their habitat. Some cutworms may still be alive and present if the field is planted quickly after ground preparation.

### Spider mites, *Tetranychus* sp.

### Damage symptoms

» Spider mites can stunt a plant's development, prevent it from blossoming, and reduce the quantity of seeds it produces.

» When mites attack young plants, they can do a lot of damage.

» During the dry season, mite damage could be even worse.

### Pest identification

» Adult females only grow to be about 0.6 mm in length.

» It's the male that's smaller in stature.

### Management

» You shouldn't put plants near infected fields.

» In order to prevent spider mite infestations, you should limit your usage of broad- spectrum insecticides, especially Pyrethroids.

» To remove mites and their webs, water plants vigorously or use overhead watering.

## DISEASES

### Damping-off, *Pythium aphanidermatum*

### Symptoms

» The seeds failed to germinate properly.

» The young plants can't hold themselves up.

» Tissue lesions ranging in colour from brown to black along the plant's soil border.

» Pre-emergence The term "damping off" refers to the deterioration of a growing seed or the demise of a seedling prior to its breaking through the soil.

» When the seedlings have emerged from the soil, yet while they are still young and vulnerable, a process known as post-emergence damping-off begins.

» The damaged plants may experience root death, and their lower stems and leaves wilt and die from waterlogging.

## Favorable conditions

» Fungi flourish in damp, warm environments.

» High soil water content and low soil temperatures promote the disease.

» Planting densely without adequate aeration promotes the spread of disease.

## Survival and spread

» Fungi can live in damp soils for years, even if no host plants are present, and they can spread through the decomposition of plant matter.

» Rainwater, runoff water (such as during irrigation), and the transfer of infected plant parts, soil, and equipment are all additional vectors for the dissemination of fungal spores.

## Management

» Plant only disease-free seeds.

» Don't bury the seeds too far underground.

» Don't overwater the plants.

» To improve airflow to the young plants, avoid planting too densely.

» Maintain proper drainage.

» Spray with Dithiocarbamates

## *Choanephora* blight, *Choanephora cucurbitarum*

## Symptoms

» Bark-like lesions dripping with water.

» Fungal spores cause lesions to take on a hairy appearance.

» Perhaps results in leaf drop.

» The leaves grow large, bright, spherical lesions surrounded by black, concentrically arranged rings that are the pycnidia.

» Bony tissue is a common result of tissue death.

» The blighted, curled shape is caused by the pathogen only infecting the tips of the new shoots.

**Favorable conditions**

» Insects and human-made damage are the most common causes of fungal infection in plants.

» An increase in disease transmission occurs in hot and humid climates

**Survival and spread**

» Disease can be transmitted through the air or by planting diseased seeds.

**Management**

» Maintain proper field hygiene.

» Different cultivars of plants that are less susceptible to contamination.

» Only plant seeds that have been verified as being of the highest quality.

» If you want to maximize yield, spacing your plants out is essential.

» When fungus is detected, Copper fungicides should be used to treat the problem.

**Anthracnose,** *Colletotrichum gloeosporioides*

**Symptoms**

» There are first tiny deep necrotic lesions encircled by a yellow halo on the leaves.

» Leaves and branches perish as the lesions spread.

**Survival and spread**

» This fungus is spread by seeds and can be found in contaminated agricultural residue.

**Management**

» Care should be taken to avoid harming plants and open wounds that can be used by disease-causing organisms to invade.

» Use disease free seeds.

# References

Aderolu, I. A., Omooloye, A. A., & Okelana, F. A. (2013). Occurrence, abundance and control of the major insect pests associated with amaranths in Ibadan, Nigeria. *Entomol Ornithol Herpetol*, *2*(112), 2161-0983.

Aragón-García, A., Pérez Torres, B. C., Damián-Huato, M. A., Huerta-Lara, M., Sáenz de Cabezón, F. J., Perez-Moreno, I., ... & Lopez Olguín, J. F. (2011). Insect occurrence and losses due to phytophagous species in the amaranth Amaranthus hypocondriacus L. crop in Puebla, Mexico.

Blodgett, J. T., & Swart, W. J. (2002). Infection, colonization, and disease of Amaranthus hybridus leaves by the Alternaria tenuissima group. *Plant Disease*, *86*(11), 1199-1205.

Espitia, E. (1992). Amaranth germplasm development and agronomic studies in Mexico. *Food Reviews International*, *8*(1), 71-86.

Mureithi, D. M., Fiaboe, K. K. M., Ekesi, S., & Meyhöfer, R. (2017). Important arthropod pests on leafy amaranth (Amaranthus viridis, A. tricolor and A. blitum) and broad-leafed African nightshade (Solanum scabrum) with a special focus on host.

Seni, A. (2018). Insect pests of amaranthus and their management. *International Journal of Environment, Agriculture and Biotechnology*, *3*(3), 1100-1103.

Smith, J. D., Dinssa, F. F., Anderson, R. S., Su, F. C., & Srinivasan, R. (2018). Identification of major insect pests of Amaranthus spp. and germplasm screening for insect resistance in Tanzania. *International Journal of Tropical Insect Science*, *38*, 261-273.

# 3

# Beet Root

## INSECT PESTS

### Webworm, *Spoladea recurvalis*

### Damage symptoms

- » Larvae typically target leaves, but they can also harm flowers and pods by weaving them together and feasting within.
- » Insect larvae eat holes in leaves as they munch on the underside.
- » The entire forest may be decimated.
- » Beet growth is slowed by defoliation.

### Pest identification

- » The full-grown larvae can measure up to 25 mm in length; they are bluish-green in colour with a dark line running down the back; and finally, they turn a rosy color just before they pupate.
- » The adult moth is about 10 mm in length, with a wingspan of 22 to 24 mm, and has distinctive white bands across the abdomen and wings.

### Spread

- » The moth is well-known for its epic journeys, which it takes to spread the disease.
- » The international trade of plants may potentially contribute to the spread of disease.

## Management

- » Make sure the ground is cultivated and clear of weeds before you plant anything. Elimination of invasive Chenopodiaceae and Amaranthaceae species in and around croplands.

- » Get rid of the caterpillars by snipping off the leaves with the webs.

- » Spraying Neem soap on the rolled-up leaves will allow you to access the caterpillars hiding inside.

- » To be effective, pyrethrum spray or dust must be sprayed to caterpillars before they have developed significant webbing.

- » The mortality rate against third-instar larvae was quite high for *Bacillus thuringiensis* sub sp. *kurstaki.*

- » *Paecilomyces farinosus* could be used as a biocontrol agent.

- » After harvesting, collect crop trash and dispose of it.

## Aphids, *Aphis fabae*

### Damage symptoms

- » Aphids are pests that feed on plant juices, resulting in distorted, yellow leaves and other plant damage.

- » Beetroot leaves will curl up and new shoots will develop abnormally.

- » Beet leaves get bloated, roll, and stop growing.

- » Neither the root development nor the sugar content are particularly good.

- » Flowers that have been harmed have a reduced ability to produce seeds.

- » Honeydew, a chemical they secrete, causes the leaves to become sticky.

- » Sooty mould grow on the leftovers and honeydew is generated.

### Pest identification

- » The body of a bean aphid is a deep olive green to black colour, while its legs are a lighter shade of green.

### Favorable conditions

- » Aphid damage is aggravated by hot, dry weather.

### Management

- » It is imperative that infected crops be destroyed soon following harvest.

» Overwintering populations may be eliminated if the weed hosts are destroyed late in the year.

» Removing infected plants from the wild as soon as possible may help prevent the spread of aphid-borne diseases.

» The crop can be shielded by row coverings.

» Insect-killing soaps should be used. Neem soap 1% spray.

» A mixture of carbaryl, diazinon, malathion, and rotenone should be sprayed.

» Several different kinds of predatory and parasitic insects, as well as a number of fungi, prey on aphids (*Beauveria bassiana*).

» Bugs like ladybirds and lace wings prey on other insects.

» *Lysiphlebus testaceipes* is a particularly successful parasitic wasp that attacks the bean aphid.

## Leaf miner (Mangold fly), *Pegomya hyoscyami*

### Damage symptoms

» The situation is the worse for young plants.

» White maggots of various sizes can be seen feeding between the leaves.

» Trails of damage can be seen snaking through the leaf tissue.

» Large, light-colored blotched patches may form as mines merge and expand.

» The holes made by the larvae are visible on the leaf.

» The plant's leaves begin to brown, and its development slows.

» Leaves with blisters and a light brown coloration, and eventually the leaves themselves will become brown and fall off.

### Pest identification

» Eggs come in an oval shape and have a pure white colour.

» Larvae are tiny, legless maggots that only grow to be around 1.25 cms long.

» The adults of this species are tiny flies with black and yellow patterns, including a white region in front of each eye.

### Favorable conditions

» Young beet seedlings are more easily damaged than mature beets.

## Survival

» Stay dormant as a pupa underground over the winter.

## Management

» At the first sign of infection, remove and destroy the afflicted leaves.

» If the plant's presence is hindering growth, it should be uprooted and destroyed.

» Fly protection might be achieved with floating row coverings.

» Weeds that serve as hosts, such as lambs quarter, must be eradicated.

» Spray Dipel or Abamectin.

» Parasitic wasps of the genera *Diglyphus*, *Opius*, and *Dacnusa* prey on leaf miner larvae.

## Swift moth, *Korscheltellus (Hepialus) lupulina*

### Damage symptoms

» Underground, larvae eat the plant's roots.

» Older larvae feed on larger roots, stolons, and the lower parts of plant stems, whereas younger larvae consume just rootlets.

» An initial symptom is a plant that appears ill and isn't growing as it should.

### Pest identification

» Larvae resemble maggots in coloration and can grow to be up to 50 mm in length; they are white and opaque overall, but their heads are a rusty reddish-brown. It has a dark brown peak and a reddish-brown prothoracic plate.

» The wingspan of a female ghost moth is 50–70 mm. The forewings are a yellowish buff colour with deeper linear patterns, while the hindwings are a dark brown. The average wing span of a male is 46-50 mm, and his wings will be either white or silver.

### Survival

» The larvae survive the winter.

### Management

» Removing diseased plants exposes a new growing space.

**Leaf hopper,** *Circulifer tenellus*

**Damage symptoms**

» Adults and juveniles of this yellowish green, wedge-shaped, winged insect feed on plant leaves by puncturing them and sucking out the sap.

» By draining the sap from trees, they can cause serious damage.

» In the case of severe infestations, leaves may become wrinkled and curled.

» Hopper burn, speckling, and yellowing of afflicted leaves are all possible outcomes of an infestation of this pest.

» They're not just a vector for pests; they can also transmit disease. Mostly, they are significant because of their role in spreading the curly top virus.

**Pest identification**

» Bugs that resemble miniature torpedoes and move incredibly quickly.

» Nymphs look like grown-ups, but they're much smaller and don't have wings.

» Less than a dime in size, with a wedge form and a colour range from light green to brown.

**Survival and spread**

» Feeds throughout the winter on Russian thistle and other plants found in weedy or waste regions and rangeland.

» Late in the spring, the leafhopper will travel to areas of the country where beets and other sensitive crops are grown.

**Management**

» Remove overgrown vegetation from areas where pests may be hiding.

» Spray Malathion, or Diazinon, or Sevin.

» At the first sign of an infestation, start spraying with Beauveria bassiana.

» Neem soap 1% spray.

# DISEASES

**Damping-off,** *Rhizoctonia solani*

**Symptoms**

» When seeds or sprouting seedlings are stifled before they have a chance to emerge, this phenomenon is known as pre-emergence damping-off.

» Plants may experience post-emergence damping-off after they have emerged from the soil.

» Seedlings wither and die due to taproot or rootlet degradation.

» When infected with R. solani, a plant will develop a waterlogged, sunken lesion at ground level.

» The growth of the plants that manage to survive are stunted, and the damaged areas generally develop unevenly.

## Favorable conditions

» It is easier for diseases to spread in hot and damp conditions.

» Infection can occur at temperatures as low as 15 °C, while the optimal range is between 20 and 30 °C.

» Soils with a moisture-holding capacity of 25% are prone to disease development, and excessive soil moisture exacerbates the severity of existing diseases.

## Survival and spread

» Hyphal fragments, sclerotia, or bulbils survive the winter in the soil and plant waste.

» The pathogen can spread via wind, irrigated water, or the shifting of infected soil.

## Management

» Try to choose a spot with good drainage to plant your seeds.

» To avoid or avoid disease problems caused by any pathogen, planting early in cool soils is essential. If you can get your seedlings established and developing vigorously before the soils begin to warm, they will be better able to withstand infection later on.

» Hold off planting until the soil has warmed up.

» Fumigating the soil to grow seedlings.

» Seeds are treated with either 0.070 ounces of Mycostop per pound of seed, or 0.05 to 0.08 ounces of Mycostop Mix.

» Use SoilGard 12G at a rate of 0.5 to 10 lb/A every four weeks to saturate the soil and prevent weed growth.

» When used as a seed treatment, penthiopyrad [a new succinate dehydrogenase inhibitor (SDHI) fungicide] proved highly efficient against *R. solani*.

» Reducing initial pathogen inoculum can also be accomplished through crop rotation that includes cereals (such as wheat, barley, maize, grass, and grain).

» The seedling blight disease can be effectively controlled by spraying 0.2% Mancozeb.

» *Streptomyces distaticus* successfully reduced the incidence of beet root blight in seedlings.

## Leaf spot, *Cercospora beticola*

## Symptoms

» Many tiny circular spots (up to 2 mm in diameter) appear on older leaves as a result of a pathogen.

» Leaves with brown or grey specks encircled by reddish-purple haloes.

» Dead, necrotic leaves are yellow or brown.

» Brown in the centre, the area will eventually turn grey with a red margin and might fall out, giving the appearance of a shot hole.

» The spots grow in size, stay round or oblong, and eventually cause serious leaf loss (defoliation).

» Infection and subsequent leaf loss typically start at the plant's lower leaves before spreading upward to the younger leaves.

## Favorable conditions

» Infections spread more easily when the weather is hot and muggy.

» When nighttime temperatures are over 16°C and relative humidity is above 90%, spore germination and infection proceed at a rapid pace between 25 and 35°C.

## Survival and spread

» Sclerotia, the fungal reproductive structures, survive the winter on diseased leaves.

» Sclerotia have a two-year lifespan in soil.

» Crop leftovers, diseased seeds, and infected wild weed hosts can all serve as overwintering sites for fungi.

» The spores of pathogens are dispersed by wind, rain, and raging waters.

## Management

» Only plant verified disease-free seeds.

» Keep enough room between plants so that air can flow freely.

» Destroy any volunteer plants that may have been contaminated and any crop debris that may have been left behind.

» Follow crop rotation every two to three years.

» In the early stages of disease outbreak, spraying with a 1%

» Bordeaux mixture is an efficient method of disease control.

» Turn the crop remnants back into the earth by ploughing them in.

## Downy mildew, *Perenospora farinose* pv. *betae*

## Symptoms

» Young leaves are turning yellow and warping.

» The pathogen invades the developing point, causing the plant to stunt its growth and develop twisted, yellow leaves that curl upwards.

» There is a dense mycelial growth, purple-grey in colour, on the upper and bottom surfaces of the leaves when the weather is damp.

» Quick drying and shrinking of affected leaves.

» Infected plants will have limited flower growth.

» The entire inflorescence is compact, and if the leaves grow too quickly, it can look like a witch's broom.

## Favorable conditions

» Weather that is cool and damp (ideally 8 °C).

» During at least six hours, the pathogen needs the leaves to be damp, and the temperature needs to be low (7 to 15 °C is ideal).

## Survival and spread

» This pathogen can survive the winter as mycelia in crop residues, seeds, infected weed hosts, and volunteer beet plants.

» The pathogen creates oospores (resting spores) with strong walls, allowing them to survive for extended periods of time either in the soil or within the remains of infected plants.

» The pathogen can spread from plant to plant by means of contaminated seeds.

» Sporangia are produced by mycelia that survive the winter and are carried by the wind to propagate the fungus.

## Management

» Only plant verified disease-free seeds.

» Effective field sanitation, crop rotation, and the use of resistant cultivars are all viable preventative options.

» Get rid of any weeds or volunteer beets that might be harbouring the disease.

» Dispose of the contaminated crop waste in a safe manner.

» Newly germinated seedlings are more resistant to disease when treated with Thiram (2.5-3.0 g/kg of seed).

» Tri-weekly 15-day intervals of Dithane Z-78 (0.3% dilution) spraying.

## Rust, *Uromyces betae*

## Symptoms

» Pustules of orange rust, 1–3 mm in diameter, form on the upper and lower surfaces of the leaf.

» Most notably, when larger pustules form in clusters or rings, necrotic tissue forms around them.

» It is possible for pustules to cover an entire leaf, and the orange spores that are formed inside of them will spread across the leaf's surface.

» Leaves with a severe infection become yellow and die.

» In severe cases, the disease can cause early leaf death, which in turn reduces crop output and quality.

## Favorable conditions

» Thrives in humid, warm (15 to 22 °C) temperatures.

» It is recommended that uredinospore germination take place between 10 and 22°C.

## Survival and spread

» The pathogen survives the winter in seed crops, wild beets, and weeds that are sensitive to its infection.

» Sometimes fungi can spread via their seeds.

» Uredinospores are the major vector for the disease's dissemination.

» Wind carries uredinospores.

» Because they may last for up to a year in the soil and feed off of crop wastes, teliospores are also useful as long-term overwintering structures.

## Management

» Collect contaminated crop wastes and dispose of them properly.

» Volunteer beet plants and weeds that may be easily killed by pulling or digging them up should be eliminated.

» To avoid introducing hosts into your crop rotation, you should use crop rotation with non-hosts.

» Make sure there is enough space between plants for air to flow freely.

» Effective fungicides applied in a timely manner.

## Powdery mildew, *Erysiphe betae*

## Symptoms

» Spreading, microscopic, whitish, fungal growth on the surface.

» Older leaves are more likely to be affected by fungus, and this growth might appear on either the upper or lower leaf surfaces.

» Affected leaves may eventually be completely blanketed in white powdery fungal growth as the disease progresses.

» Infected older leaves will turn yellow and may die.

» Younger leaves are susceptible to infection if the disease is particularly bad.

» In extreme cases, disease might cause a 25 % drop in harvest success.

» It's possible that beet roots with severely damaged leaves won't sell.

## Favorable conditions

» Condition caused by the dry climate.

» There is no need for free water for germination to take place because spores contain so much of it (60%).

» A temperature of 25°C and a relative humidity of 70 to 100% are ideal for spore germination.

## Survival and spread

» The fungus survives the winter by attaching itself to the roots of buried beet plants or other vulnerable weeds.

» In addition to ascospores, sexual fruiting structures (cleistothecia) of fungi can overwinter on crop trash, releasing them in the spring and causing diseases.

» The spores disperse in the wind.

## Management

» Any plant debris containing the pathogen must be removed from the field and destroyed.

» Plowing down and destroying agricultural remains after harvest is necessary.

» Volunteer beet plants and weeds that can be easily defeated by pulling them should be destroyed.

» Effective fungicides applied in a timely manner.

## References

Clifford, T., Howatson, G., West, D. J., & Stevenson, E. J. (2015). The potential benefits of red beetroot supplementation in health and disease. *Nutrients*, 7(4), 2801-2822.

Dos Santos Baião, D., Vieira Teixeira da Silva, D., & Margaret Flosi Paschoalin, V. (2021). A narrative review on dietary strategies to provide nitric oxide as a non-drug cardiovascular disease therapy: Beetroot formulations—A smart nutritional intervention. *Foods*, 10(4), 859.

Harveson, R. M., Hanson, L. E., & Hein, G. L. (2009). *Compendium of beet diseases and pests* (No. Ed. 2). American Phytopathological Society (APS Press).

Jones, F. G. W. (1957). Sugar beet pests. *Sugar beet pests.*, (162).

Lange, W. H. (1987). Insect pests of sugar beet. *Annual review of entomology*, 32(1), 341-360.

Martin, H. L. (2003). Management of soilborne diseases of beetroot in Australia: a review. *Australian Journal of Experimental Agriculture*, 43(11), 1281-1292.

Siddiqui, Z. A., Khan, M. R., Abd_Allah, E. F., & Parveen, A. (2019). Titanium dioxide and zinc oxide nanoparticles affect some bacterial diseases, and growth and physiological changes of beetroot. *International Journal of Vegetable Science, 25*(5), 409-430.

Sivarajah, N., & Emmanuel, C. J. (2022). Impact of leaf spot disease on beetroot production in Jaffna peninsula, Sri Lanka. *Archives of Phytopathology and Plant Protection, 55*(13), 1558-1570.

Wolf, P. F. J., & Verreet, J. A. (2002). An integrated pest management system in Germany for the control of fungal leaf diseases in sugar beet: The IPM sugar beet model. *Plant Disease, 86*(4), 336-344.

# 4 Brinjal

## INSECT PESTS

### Shoot and fruit borer, *Leucinodes orbonalis*

#### Damage symptoms

- » Caterpillars cause withering by feeding on the fragile shoots just before they bloom.

- » The mature larvae then travel to new flower buds and eventually bear fruits, polluting them with their own waste.

- » Unopened flower buds enlarge and become home to the borer during periods of high prevalence.

- » Mature larvae emerge from fruits and flower buds just before they pupate, settling on plant parts or trash to spin silken cocoons in which to develop into adults.

#### Pest identification

- » The full-grown larva ranges in size from 16 to 23 mm and is a bright pink colour with tiny hairs covering its bumpy body and a dark brown or blackish head.

- » The moth can be either white or a dirty white colour, and it has pale brown or black patches all over its dorsal (upper) body. Front wings are spotted with pink or brown and red, and the back wings are spotted with red. The female is noticeably largerand more full-figured than the male.

#### Management

- » Nylon netting can be used as a barrier between nurseries and main fields to prevent the spread of pests.

» During the pre-flowering and flowering stages, cut off and dispose of any damaged shoot tips caused by insects.

» Larvae in swollen, damaged flower buds and fruits must be destroyed routinely, ideally after each harvest.

» Pheromone lure (Lucinlure) in conjunction with portable water traps was efficient in mass trapping. This was accomplished by growing barrier crops, such as maize, around the crop.

» After the flowers have bloomed, spread 250-500 kg/ha of Neem or Pongamia cake along the ridges. Two further sessions at a 30-45 day gap are recommended.

» Every 10 days, spray NSPE 4% or Neem oil 2%.

» Neem soap (at 7.5 g/L) with either Enodsulfan 35 EC (2 ml/L) or Cypermethrin 25 EC (0.75 ml/L) can be mixed and sprayed.

» Combination of weekly shoot clippings of infested plants, followed by Neem cake application (250 kg/ha) 30 days after planting, and sprays of 4% crushed NSPE/ 1% Neem soap/ 1% Pongamia soap at 60, 75, 90, and 105 days after planting to control the borers.

» Following the application of a 1% *Bacillus thuringiensis* formulation at weekly intervals, *Trichogramma chilonis* should be released at a rate of 2,500,000/ha (50,000/release-5 times starting from blooming).

## Leafhopper, *Amrasca biguttula biguttula*

### Damage symptoms

» Nymphs and adults alike use piercing and sucking mouthparts to extract sap from the undersides of leaves.

» They inject their poisonous saliva into the plant tissues as they suck the sap, causing the plant to become yellow.

» Leaves get "hopper burn," a condition characterized by yellow spots, wrinkling, curling, bronzing, and drying, caused by the simultaneous feeding of several insects.

### Pest identification

» The nymphs, which are a drab green, seem just like the adults except that they have no wings and only slightly enlarged wing pads.

» The wings of an adult are fully formed, and there is a black speck on each of the pale green wings.

## Management

» Place yellow (570-580 nm) sticky traps around the field and observe capture rates.

» Using okra as a trap crop around eggplant fields can help keep pests at bay.

» The hopper population can be reduced by spraying the soil with NSKE 4% or Neem soap 1% every 10 days after applying Neem cake at a rate of 250 kg per hectare.

» Dimethoate 30 EC @ 2 ml/L, Imidacloprid 200 SL @ 0.3 ml/L, or Acephate 75 SP (1 g/L) sprayed at the pre-flowering stage will reduce the pest population.

» For example, parasitoids like *Anagrus flaveolus* and *Stethynium triclavatum* might be quite helpful.

» Manjari Gota, Vaishali, Mukta Kesi, Round Green, Kalyanipur T3, and the Bangladeshi variation Bagun 6 are among the cultivars rumoured to be more damage-resistant or resilient.

## Ash weevil, *Myllocerus subfasciatus*

## Damage symptoms

» The ash weevil is a pest found all throughout the world that kills ash trees.

» Adults create saw-toothed damage along leaf margins as they eat vegetation.

» Root-eating grubs.

» The affected plants wither and die.

» Wilting typically appears as bare spots on a plot at first.

» When such withering plants are uprooted, only few roots are seen.

» Scratching can be seen on the plant's roots and stem.

## Pest identification

» Grubs are tiny, apodous creatures.

» The elytra of adults might be greenish white with dark lines, brownish weevil, brown with a white spot, or a tiny and pale green.

## Management

» Fifteen days after planting, spread 15 kg/ha of carbofuran 3G.

» Light traps set at a density of 1/ha will be effective in attracting and killing.

» Protect *Pristomerus testaceus* and *Pristomerus euzopherae*, two parasitoids that feed on larvae.

» From one month after planting, spray with a single pesticide every 15 days (Carbaryl 50 WP 2 kg + Wettable Sulfur 50 WP 2 kg, or any combination).

» NSKE 5%, Azadirachtin 1%, Fenpropathrin 30 EC 250-340 ml, or Thiodicarb 75 WP 625-1000 g.

» Neem cake, at a rate of 250 (kg/ha), was incorporated into the soil prior to the planting of seedlings. A Neem cake treatment should be repeated once or twice at 30 and 60 DAP.

## Epilachna beetle, *Epilachna vigintioctopunctata*

### Damage symptoms

» The beetle grubs and adults both have mouthparts adapted for chewing.

» To extract chlorophyll, they peel the outermost cuticle of the leaves.

» The resulting window looks like a standard ladder after being fed.

» The leaves will get pockmarked as the windows dry and fall off.

» When several windows join together, it causes skeletonization, which looks like a papery structure on the leaf and indicates a serious infestation.

» The pest can also ruin your produce.

### Pest identification

» These birds lay their 10-40 yellowish, spindle-shaped eggs in groups.

» Grubs have a body covered in black spiky hairs and a creamy white or yellowish white tint.

» Although pupae and grubs look similar, pupae are often deeper in colour and can even appear yellow at times. The back of its body is covered in prickly hairs, whereas the front is not.

» Adults of this species are hemispherical beetles of a brownish or orange color, with 28 black spots across each of their forewings (elytra).

**Management**

» Adult beetles are being gathered and disposed of.

» If necessary, spray a contact pesticide, such as Carbaryl 50 WP at 3 g/l or Quinalphos 40 EC at 1.5 ml/l.

» *Pediobus foveolatus*, which caused parasitism at rates of 60-77%, successfully held it in check.

» The grubs of the epilachna beetle were completely eradicated from brinjal plants after 10 days after being sprayed with a 'Thuricide' (*Bacillus thuringiensis*) dust at a dose of 3x106 spores/mg.

» Arka Shirish, Hissar Selection 14, and Shankar Vijay are only some of the varieties that have been reported to be tolerant of, or even resistant to, Epilachna beetle, particularly *E. vigintioctopunctata*.

## Gall midge, *Asphondylia capparis*

**Damage symptoms**

» The brinjal bloom and fruit is frequently attacked by the gall midge.

» This pest's maggots feast on the developing ovary or fruit.

» If pests get to young flowers, they'll ruin the stamens and pistils, which will lead to a lack of fruit.

» The fruit becomes unmarketable because it has developed cracks along the curve's inner side after the fly has emerged.

**Management**

» This bug can be effectively eradicated by spraying 4% NSKE.

## Mealy bug, *Coccidohystrix insolita*

**Damage symptoms**

» The mealybugs spread a white pile on the leaves of solitary, elderly plants.

» When insects feed on plants, the plants become drained of colour and vitality.

**Management**

» Mealy insect infestations can be effectively sprayed with either 0.4% Monocrotophos or 0.15 % Malathion.

» The release of the predator *Cryptolaemus motrouzieri* also helped keep the population under control.

## Whitefly, *Bemisia tabaci*

### Damage symptoms

» Affected leaves become yellow, roll inward, and fall off the plant.

» Defoliation and slow growth are symptoms observed in plants.

» Whiteflies are sap-sucking insects that leave behind sweet honeydew on plants' foliage and fruit.

» The honeydew becomes infested with sooty mould fungus, decreasing both fruit quality and quantity.

### Pest identification

» The adults resemble small white moths; their entire bodies and wings are covered with a white waxy bloom.

» Puparia and nymphs, both oval and scale-like in appearance, are found on the undersides of leaves.

» The pear-shaped eggs are pale in colour and have tiny stalks.

### Management

» To keep track of whiteflies, set up yellow sticky traps.

» Whiteflies on brinjal could be effectively controlled by spraying them with a 1% solution of Neem or Pongamia soap.

» Whitefly on brinjal was reduced by 80.29 and 74.37 % after spraying with Thiamethoxam 35 g a.i. ha$^{-1}$ and Acetamaprid 20 SP g a.i. ha$^{-1}$, respectively.

» For whiteflies specifically, the most effective treatment was Imidacloprid 17.8 SL @ 50 g a.i. ha-1, which resulted in the lowest whitefly population per plant (1.55), the greatest decrease of whiteflies (83.15 %), and the highest marketable fruit output per hectare (14.65 tonnes per hectare).

» The natural population of whiteflies can be kept in check by protecting parasitoids such the *Encarsia* sp., *Eretmocerus* sp., and *Chrysocharis pentheus* (nymphal) species.

» Protect the insects that eat whiteflies, like ladybirds (*Coccinia septumpunctata*), lacewings (*Chrysoperla zastrowi sellemi*), mirid bugs (*Dicyphus hesperus*), dragonflies, spiders, and big-eyed bugs (Heteroptera).

**Aphid,** *Aphis gossypii*

**Damage symptoms**

- » Sap is sucked from leaves by both juveniles and adults.
- » Diseased leaves become yellow, wrinkled, misshapen, and eventually die.
- » Honeydew is secreted by the insect and then a fungus grows on it, quickly covering the plant in sooty mould and preventing it from photosynthesis.
- » Because of this, the plant's growth is inhibited and the harvest is diminished.
- » Necrotic patches on leaves and/or reduced growth of new shoots may be the result of a severe aphid infestation.
- » They also serve as a vehicle for the spread of many viruses.

**Pest identification**

- » The adult aphid is just around 1 mm in length and has two projections called cornicles on the dorsal side of the abdomen.
- » Aphids can be easily identified by their cornicles, which are tubular structures that extend behind the insect's body; aphids also typically do not react quickly to being disturbed.

**Management**

- » If an aphid infestation is only present on a few numbers of leaves or shoots, then control can be achieved by removing those parts of the plant that have been infested.
- » Before planting, make sure the transplants are free of aphids.
- » Aphids can be deterred from plants by using reflective mulches, such as silver coloured plastic.
- » Aphids on sturdy plants can be dislodged using a vigorous water jet.
- » The most effective way of control is the use of insecticidal soaps or oils like Neem or Canola oil.
- » Green lacewing (*Chrysoperla carnea*) first-instar larvae should be released at a rate of 4,000 larvae per acre.
- » Save parasitoids like *Aphidius colemani, Diaeretiella* species, *Aphelinus*, etc.
- » Retain predators including ladybird beetles (*Coccinella septumpunctata* and *Menochilus sexmaculatus*), lizards, and snakes.

» Much less aphids were found in brinjal after a bare-root dipping treatment with 3% Neem oil was followed by a spray with 5% NSKE.

» The pest can be effectively controlled by spraying it with either 0.05% Endosulfan, 0.02% Phosphamidon, 0.03% Dimethoate, Methyl demeton, or Thiometon.

» You can apply Betacyfluthrin 8.49% + Imidacloprid 19.81% OD at 70-80 ml in 200 ml of water/acre, Fenvalerate 20% EC at 150-200 ml in 240-320 ml of water/acre, Phorate 10% CG at 6000 g/acre, Phosphamidon 40% SL at 250-300 ml in 300 ml of water/acre, Thiometon 25%

## Red spider mites, *Tetranychus urticae*

## Damage symptoms

» Typically, spider mites use their long, needle-like mouthparts to eat the contents of the leaves' cells.

» When the leaves' chlorophyll levels drop, white or yellow spots appear on them.

» When an infestation is bad enough, the leaves wither and fall off.

» In extreme cases, the mites will also generate webbing on the leaf surfaces.

» As mite populations get too dense, they migrate to the plant's tip, where they construct a ball out of their silken threads. This ball is then carried by the wind to nearby leaves or plants.

## Pest identification

» Eggs tend to be spherical and colourless or very pale.

» Adults range from green to greenish yellow to brown to orange red, and always have two dark spots somewhere on their little bodies.

## Management

» Rapidly removing and disposing of contaminated plants is essential for stopping the spread of a disease.

» Sprays of Wettable Sulfur 75 WP @ 3g/L or Dicofol 18.5 EC @ 2.5 ml, both of which are acaricides.

» Neem soap/Pongamia soap 1% is a chemical-free alternative to acaricides.

» It is imperative that the underside of the leaves be thoroughly sprayed.

» Use Tetracycline antibiotics to treat diseased plants.

» Several predators, such as *Stethorus* spp., *Oligota* spp., *Anthrocnodax occidentalis*, *Feltiella minuta*, etc., have been documented.

» Spider mites can be managed with the help of predatory mites like *Phytoseiulus persimilis*, *Amblyseius womersleyi*, and A. fallacies.

» Moreover, green lacewings (*Mallada basalis* and *Chrysoperla carnea*) are efficient. *C. carnea* grubs, especially those in their third instar, can swallow as many as 25-30 adult spider mites in a single day.

» According to reports, some plant varieties can tolerate or even resist the disease. They include the Pusa Purple Long, Pusa Purple Round, Pusa Purple Cluster, Nurki, Hisar Shyamal, and H-10.

# DISEASES

### Phomopsis blight, *Phomopsis vexans*

### Symptoms

» Damping-off is the symptom of the disease in the seed beds.

» As the plants are transplanted, they develop distinct circular spots on the leaves, ranging in colour from grey to brown with a lighter centre.

» Symptoms include the yellowing and eventual death of the afflicted leaves.

» Stem lesions feature a dark brown outside and a grey inside.

» Stem base constriction or dry grey rot is a symptom of stem base disease.

» Initially appearing as isolated pale depressions, the fruit's surface may eventually become completely covered in these lesions.

» The rotting occurs in the fruit's centre.

» When the calyx is broken and the fungus is able to infiltrate the fruit, the entire fruit will dry rot and become a mummy.

### Favorable conditions

» Fast growth of the fungus is facilitated by both moisture and warmth.

### Survival and spread

» The fungus can spread from crop to crop by means of infected seeds, as well as by remaining dormant in the soil and on plant waste from previous crop seasons.

» Easily disseminated by means of farm machinery and irrigation water.

## Management

» It's important to plant only disease-free seeds.

» Seeds should only be collected from plants that are in good health and from unblemished fruit.

» It's recommended to do deep ploughing in the summer.

» Seeds need to be soaked in water heated to 50°C for 30 minutes.

» Protecting seedlings in the nursery phase using Thiram (2 g/kg seed) therapy.

» Since the pathogen only infects brinjal, rotating crops every three years should help control the disease.

» Gather the diseased crop waste and burn it out there in the field.

» Disease in the field can be efficiently controlled by spraying with either Zineb (Dithane Z-78, 0.2%) or Bordeaux mixture (1%).

» Spray either Foltaf or Copper oxychloride at a dosage of 0.3%.

## Damping-off, *Phytophthora, Rhizoctonia* or *Pythium* sp.

## Symptoms

» Before they even have a chance to emerge from the soil, pre-emergence damping off kills off the seedlings.

» Post-emergence damping off is defined by the collapse of infected seedlings at any moment after they emerge from the soil and before the stem has stiffened enough to resist invasion.

» As the disease progresses, the plant's stems eventually distorted and the plant dies.

## Management

» Adjusting the height of cribs and changing tables.

» Baby cribs with draining raised beds.

» Overcrowding can be avoided by scattering the seeds.

» Moderate amounts of water applied frequently.

» *Trichoderma harzianum, Penicillium oxalicum, Pseudomonas cepacia*, and *Pseudomonas fluorescens* seed coating.

» Swap out your annual Solanaceous crop for anything else.

### *Cercospora* leaf spot, *Cercospora solani -melongenae, C. solani*

#### Symptoms

» Chlorotic lesions on the leaves are angular or irregular in appearance and eventually develop greyish brown with abundant sporulation at their centres.

» Premature leaf drop from severe infection reduces fruit production.

#### Survival and spread

» Infectious conidia can float through the air and spread the disease.

#### Management

» Carbendazim sprays with a concentration of 0.1% are useful for disease control.

» Planting disease-resistant cultivars is an effective strategy for disease control. The 'Pant Samrat' cultivar has shown to be resistant to both of these leaf spots.

### Alternaria leaf spot, *Alternaria melongenae, A. solani*

#### Symptoms

» There are round, dark- to light-brown, ring-shaped markings on the leaves.

» Leaf spots can be isolated or widespread, and their progression from yellow to brown and eventual exfoliation is characteristic of a fungus.

» Spots can be anywhere from 4 mm to 8 mm in diameter and can sometimes join together to cover vast portions of the leaf.

» Leaf patches develop cracks.

» All of the plant's leaves could fall off if the disease is bad enough, and the fruit might burn in the sun if it's exposed to the elements.

» Little fruits are covered in concentric dark brown and sunken patches that eventually combine into enormous deeply seated blotches.

» Infected fruit wilts and falls off the tree early.

#### Survival and spread

» The conidia in the air are what actually transmit the disease.

## Management

> » It's time to cut off the diseased branches and burn them.

> » Plants should be thinned out to improve ventilation. Wet environments are ideal for the infection.

> » Spraying 1 % Bordeaux mixture, 2 grammes of copper oxychloride, or 2.5 grammes of zinc phosphide per litre of water is an excellent method of preventing and treating leaf spots.

> » For effective disease management, spray infected plants with Bavistin (0.1%).

## Collar rot, *Sclerotium rolfsii*

## Symptoms

> » Because the inoculum is carried through the soil, it primarily affects the plant's lower trunk (sclerotia).

> » The most noticeable sign is decortication.

> » The plant could die from the exposure and necrosis of its underlying tissues.

> » Mycelia and sclerotia can be observed on the stem close to the ground.

> » A lack of plant vitality, water buildup around the stem, and mechanical damage all play a role in the progression of this disease.

## Management

> » Treating seeds with a formulation of *Trichoderma viride*, at a rate of 4 g per kg of seed, will aid in disease suppression.

> » Acquiring and destroying infected plant materials.

> » Mancozeb spray, 2 g/ lit of water.

## Bacterial wilt, *Ralstonia solanacearum*

## Symptoms

> » The signs of this disease include wilting, stunting, and yellowing of the plant's foliage, followed by the plant's eventual collapse.

> » The drooping of the lower leaves is often the first sign of impending withering.

> » Infection is mainly restricted to the vascular system. The blood vessels start to turn brown.

» The infected plant portions will exude a white bacterial goo when cut and placed in clean water.

» Tissues may get discoloured and turn a yellowish brown as it spreads into the cortex and pith.

» By midday, plants begin to exhibit signs of wilting; they perk up at night, but ultimately perish.

## Survival and spread

» It has been discovered that the pathogen can survive in the soil and on the decaying plant matter for up to 10 months.

» Disseminates in farm machinery and irrigated water.

## Management

» Affected plant components should be collected and burned, and fields should be kept clean.

» When root-knot nematodes are present, the disease becomes more widespread.

» Hence, reducing these nematodes will slow the progression of the disease.

» Cowpea-maize-cabbage, okra-maize, maize-maize, and finger millet-brinjal (Pusa Purple Cluster)-French bean are all examples of crop rotations.

» The disease can be kept under control by spraying it three times every 10 days with either a 2% Bordeaux mixture or Agrimycin-100 at 100 ppm (0.1g/liter).

» We observed that grafting brinjal onto a *Solanum torvum* rootstock protected the plant from wilt.

» *Pseudomonas fluorescens* establishment in the soil.

» Banaras Giant Green, Gulla, Vijay hybrid, and Pusa Purple Cluster are resistant, as are a number of Dingra Purple varieties. Tolerant 'Pant Samrat' variety.

## Mosaic, *Tobacco Mosaic Virus* (TMV)

## Symptoms

» The leaves of an infected plant are tiny, leathery, and misshapen. Few fruits develop on diseased plants.

» TMV causes noticeable mottling of leaves, which is a major symptom. The early stages of mosaic are not severe, but progress to become so later.

» Eventually, blisters will appear on the leaves as well. Those leaves that are severely diseased shrink and change shape. Prematurely infected plants continue to show signs of deformity.

» Sap, contaminated tools and clothing, dirt litter, and worker hands are all potential vectors for the spread of the tobacco mosaic virus.

» It may thrive on a wide variety of crops and weeds.

**Survival and spread**

» In soil, the virus can be found in dead plants.

» The mosaic virus is carried by insects that feed on perennial weeds and survive the winter there.

» The plant could have gotten the virus from contaminated soil, seed, starter pots, or containers.

» Similarly, the virus can be spread via cuttings or divisions from infected plants.

**Management**

» All contaminated plants must be uprooted and destroyed.

» Remove weeds and avoid planting cucumber, pepper, tobacco, or tomato anywhere near your brinjal seed beds or field.

» Preparing to work in the seed beds requires washing hands with soap and water.

» Anyone working with brinjal seedlings shouldn't smoke or chew tobacco.

**Little leaf, *Phytoplasma***

**Symptoms**

» Symptoms include a significant decrease in leaf size, a bushier look due to the clustering of many leaves on one branch, and a shortening of internodes.

» This plant has such short petioles that the leaves appear to be glued to the stem.

» These leaves are thin and yellow in colour.

» The aetiology of this lower length is unclear.

» Stem internodes are reduced in length as well.

» Despite their growth, the petioles and leaves of axillary buds remain rather short.

» This causes the plant to appear more bushy.

» In most cases, flowering does not occur, and in those where it does, the resulting blooms are small, narrow, pale green, soft, and smooth, and the resulting fruits are smaller.

» Infrequent flowering.

» Because of the disease, the plant will eventually die.

**Survival and spread**

» The virus relies on weeds as hosts to keep it alive. This pathogen can infect a wide variety of hosts.

» Leaf hoppers, *Hishimonas phycitis, Empoasca devastans*, and grafting are all vectors for the disease. In terms of vector competence, *E. devastans* does poorly.

**Management**

» Destroying infected plants helps lessen the impact of the disease.

» Soil Phorate treatment after a Tetracycline or Monocrotophos immersion for the seedlings.

» Streptocycline at 100 ppm can be sprayed on plants during the early stages of infection to prevent the spread of phytoplasma. In early seeded crops, the spraying process should be performed twice.

» Methyl demeton 25 EC, dimethoate 30 EC, or malathion 50 EC, sprayed at 2 ml/L.

» In the field, cultivars like Pusa Purple Cluster, Arka Sheel, Aushy, Manjari Gota, and Banaras Giant display moderate to high levels of resistance.

» Black Beauty, Brinjal Round, and Surati are among more cultivars discovered to be disease-resistant.

# References

Anjali, M., Singh, N. P., Mahesh, M., & Swaroop, S. (2012). Seasonal incidence and effect of abiotic factors on population dynamics of major insect pests on Brinjal crop. *Journal of environmental research and development*, 7(1A), 431-435.

Borah, N., & Saikia, D. K. (2017). Seasonal incidence of major insect pests of brinjal and their natural enemies. *Indian Journal of Entomology*, *79*(4), 449-455.

Borkakati, R. N., Venkatesh, M. R., & Saikia, D. K. (2019). Insect pests of Brinjal and their natural enemies. *Journal of Entomology and Zoology Studies*, *7*(1), 932-937.

Najar, A. G., Anwar, A., Masoodi, L., & Khar, M. S. (2011). Evaluation of native biocontrol agents against Fusarium solani f. sp. melongenae causing wilt disease of brinjal in Kashmir. *Journal of Phytology*, *3*(6).

Omprakash, S., & Raju, S. V. S. (2014). A brief review on abundance and management of major insect pests of brinjal (Solanum Melongena L.). *International Journal of Applied Biology and Pharmaceutical Technology*, *5*(1), 228-234.

Pandey, A. (2010). Studies on fungal diseases of eggplant in relation to statistical analysis and making of a disease calendar. *Recent research in Science and Technology*, *2*(9).

Patil, R. S., & Nandihalli, B. S. (2009). Seasonal incidence of mite pests on brinjal and chilli. *Karnataka Journal of Agricultural Sciences*, *22*(3), 729-731.

Sharma, D. K., Sharma, N., & Rana, S. (2013). Seed-borne diseases of brinjal (Solanum melongena L.) and their control measures: A review. *International Journal of Bioassays*.

Sidhu, A. S., & Dhatt, A. S. (2006, December). Current status of brinjal research in India. In *I International Conference on Indigenous Vegetables and Legumes. Prospectus for Fighting Poverty, Hunger and Malnutrition 752* (pp. 243-248).

Singh, B. K., Singh, S. S. B. K., & Yadav, S. M. (2014). Some important plant pathogenic disease of brinjal (Solanum melongena L.) and their management. *Plant Pathology Journal (Faisalabad)*, *13*(3), 208-213.

# 5 Button Mushroom, Paddy Straw Mushroom and Oyster Mushroom

## INSECT PESTS

### Phorid fly, *Megaselia halterata*

#### Damage symptoms

- » Larvae of phorids can only survive by feeding on mycelia.
- » They primarily consume the expanding hyphal tips of mushroom mycelium, and only occasionally the fruiting body.

#### Pest identification

- » Although adult phorids are smaller than sciarids (about 2-3 mm), they are far more robust insects. In terms of appearance, both male and female flies are similarly dark and have a hump on their backs.
- » Maggot-like larvae are a dingy off-white colour and lack a prominent head capsule.
- » The front is pointed, while the back is flat and somewhat rounded.
- » Boat-shaped eggs may also have a palisade of flat platelets like a gun barrel protecting the respiratory plastron from the rest of the embryo.

#### Management

- » Fast decomposition and maximum warmth production.
- » Filters and secure doors are used to keep adults out, particularly in the first three weeks after spawning.
- » Cuttings should be discarded far away from the growing facility.
- » Pyrethrums are effective in killing adult flies, including those of the sciarid and phorid families.

» Megaselia populations can be controlled with neem and other plant-based botanicals.

» Some evidence suggests that the juvenile hormone mimic methoprene is effective against them.

» *Bacillus thuringiensis* var. *israelensis* is a microorganism that has proven effective in controlling pests in field conditions.

## Sciarid fly, *Lycoriella castanescens, L. ingénue*

### Damage symptoms

» The larvae of this fly feed on a wide variety of organic matter, including mushroom compost, mycelia, spawn grains, mushroom primordia (pins), and carpophores.

» While the mushroom primordia is very young, up to around 1.5 cm in diameter, the larvae can eat all of the tissue inside.

» The mushrooms will have a glossy, light brown appearance; the tiny carpophore may be fully pierced; and the tissues may crumble when harvested.

» Carpophores that are particularly large when attacked will have black necrotic regions in the stipe where the larvae have been feeding.

» Compost-eating sciarids have a strong preference for compost that has not been colonised by *Agaricus mycelium.*

» Some larvae, instead of boring into the stipe, feed on the mycelia near the stem's base, stunting the mushroom's growth.

» Mites and diseases, such as Verticillium, can be spread by flies because the spores of these diseases are sticky and stick to the bodies of flies.

### Pest identification

» The eggs range in shape from spherical to oval and are 0.7 x 0.3 mm in dimensions.

» Larvae of the sciarid family are white, elongated maggots with a shiny black head.

» Mature sciarids are little (about 3–4 mm in length), fragile, two-winged flies that are mostly dark grey or black and have huge compound eyes and long, threadlike antennae. Female are larger than men because their tummies expand to accommodate egg production.

## Management

» Fast decomposition and maximum warmth production.

» Filters and locked doors keep adults out. Doorways should not be kept open to facilitate air flow.

» Disposing of crop waste away from the farm.

» Treatment options for sciarid larvae include the growth regulator Dimilin and the insect parasitic nematode Nemasys M.

» Pyrethrums are effective in killing adult flies, including those of the sciarid and phorid families.

» Successful predatory mites against sciarids include *Stratiolaelaps* (Hypoaspis) *miles*, *Geolaelaps* (Hypoaspis) *aculeifer*, and *Stratiolaelaps* (Hypoaspis) *scimitus*.

» The entomopathogenic nematodes, especially *Steinernema feltiae*, and the insect growth regulator Methoprene are more effective against older larvae, whereas *Bacillus thuringiensis* var. *israelensis* and Diflubenzuron are more effective against younger ones.

## Cecid fly, *Heteropeza pygmaea*

### Damage symptoms

» The stipe is grooved vertically by larvae as they consume mycelium.

» Outside of the stipe, where the stipe meets the gills, is where they feed.

» Brown discoloured stripes appear on the stipe and gills when bacteria is present on the skin.

» Little pustules of a dark fluid develop as delicate gill tissues deteriorate.

» Juvenile larvae also consume the sap that is secreted.

» Their presence causes condensation, and they disperse the browning-causing bacteria.

» Direct yield loss or a reduction in the amount of fresh or processed food that can be sold as a result of their presence are also possible outcomes.

» When it comes to mushrooms, ceci can be responsible for a spoiling rate of up to 50 %.

### Pest identification

» Maggot-like larvae can be either white or orange in appearance.

» Little as they are, adult flies are rarely seen, but paedogenic larvae are common.

## Management

» The barbecue went well, and the place was cleaned and disinfected quickly. Compost temperatures must reach 55 °C for 5 hours or 60 °C for 3 hours in order to destroy ceci and their larvae.

» Tight cleanliness standards. Filters and locked doors keep adults out. Doorways should not be kept ajar to facilitate air flow.

» Fast decomposition and maximum warmth production.

» Methoprene and Permethrin had a sub-lethal effect on both Mycophila and Heteropeza (in vitro), decreasing larval fecundity and lengthening the paedogenic life cycle, which could be useful in managing cecids.

» Reduce the number of developing larvae with the use of diazinon or lindane.

## Sphaerocid fly, *Pullimosina heteroneura*

## Damage symptoms

» Large populations of sphaerocerids can cause significant problems.

» Similar issues arise during infestations, including disease spread, damaged compost, and decreased yields.

» Compost and artificial light may attract them, as they do sciarids.

## Management

» When Nemasys M was applied to the casing, no flies ever emerged from the containers.

» Nemasys M gives users more granular control over their Sphaerocerid populations than any other chemical product before it.

## Tarsonemid mite, *Tarsonemus myceliophagus*

## Damage symptoms

» They are destructive because they eat just the hyphae of mushrooms.

» A reddish-brown discoloration will appear at the mushroom's stem base.

» In extreme cases of infection, the entire mushroom base may become separated from the developing surface.

## Pest identification

» Mites are very tiny, requiring a microscope to see them in their natural light brown colour.

## Management

» To ensure that all mites are destroyed throughout the pasteurisation process, proper composting and peak heating are required.

» It's important to maintain a clean agricultural environment, especially when harvesting crops.

## Red pepper mite, *Pygmephorus* spp.

## Damage symptoms

» Their motion on the surface of a mushroom cap or casing causes the accumulations to look reddish brown to the naked eye.

» Mites consume mycotoxins produced by rival fungus (generally on *Trichoderma*).

» While pepper mites aren't strictly speaking a mushroom problem, they can help a green mould infestation grow throughout a room.

» Despite the fact that red pepper mites do not directly harm farmed mushrooms, their presence often contributes to a drop in marketable yield.

## Pest identification

» Mites are 0.25 mm in length, flattened, and a yellowish-brown tint.

## Spread

» Flies riding on the clothes of harvesters can help spread the disease in this way. The watering operation spreads the mites to adjacent shelves and even the floor due to the splashing water.

## Management

» For best results, use mite-free, high-quality compost.

» By properly preparing and pasteurising the compost, we can reduce the number of weed moulds and, in turn, eradicate red pepper mites.

» Use salt or disinfectant-soaked paper towels over gas burners to treat diseased areas.

» The spread of the mites and moulds can be limited by taking care with fly management, shared equipment transported between production rooms, shared clothing and footwear, shared picking baskets, and shared sales containers (also known as punnets).

## Tyroglyphid mites, *Tyrophagus* spp.

### Damage symptoms

» Caps and stems with tiny pits are a favourite snack.

» Bacterial decay then occurs in these craters.

» As a result, the skin collapses, exposing the pit beneath.

» It is possible that tyloglyphids will eat mushroom mycelium, reducing yields.

### Management

» In areas with effective composting and peak heating, mites shouldn't be a problem.

» A farm shouldn't let organic material pile up because it serves as a haven for pests like mites.

# DISEASES

## Lipstick mold, *Sporendonema purpurescens*

### Symptoms

» This sickness first manifests as a white crystalline-like mould in spawning compost, making it difficult to tell apart from the spawn itself.

» The mould spores progress from white to pink to cherry red to dull orange or buff as they grow.

» The white mycelial growth typically occurs in the crevices of the casing and has the ability to proliferate in healthy compost.

» Lipstick mould is often a secondary disease in crops where a virus has caused significant damage.

### Spread

» Primary inoculum can be obtained from soil, casing mixture, or used compost.

» Splashes of water or harvesters spread it even further.

» Lipstick fungus is thought to be present in the litter.

**Management**

» Cleanliness is important.

» The pathogen can be removed from the compost by pasteurizing and curing it properly.

## Wet bubble, *Mycogone perniciosa*

### Symptoms

» Mushrooms are prone to developing distorted masses (Sclerodermoid masses) of fake tissue, which are initially white and fluffy but turn brown as they age and die.

» At high moisture/RH circumstances, the surface of the undifferentiated tissue develops tiny drops of liquid that range in colour from amber to dark brown.

» Mycelial proliferation on the surface of the casing may manifest as tiny, amorphous white patches.

» Any damage to ripe fruit will cause the base of the stem to become brown.

» Cap manifests its disease as a patch of short, stunted gills.

### Survival and spread

» Spores can live in agricultural detritus and contaminate following crops in temperate climates.

» Generally speaking, contaminated casing is the main way the disease spreads among farms.

» Aleurospores and conidia can both propagate secondary through splashing in water.

### Management

» Regular disinfection of equipment and employee uniforms is necessary to stop the spread of inoculum.

» Do not soak your crops by overwatering the beds where they will be grown.

» Under a bright light, carefully pick out any diseased mushrooms.

» Regular usage of gloves and disinfectants are recommended for pickers.

» Before harvesting, make sure all sick mushrooms have been removed.

» You can stop the spread of the disease by covering the infected mushroom with common salt or pots (be sure to push the pots all the way into the casing).

» If you want to water your mushrooms, wait till after you've gotten rid of those that appear to be sick.

» Benomyl, Carbendazim, and Thiobendazole, all benzimidazole fungicides, are highly effective at a 0.05% concentration.

» Good management is also provided by prochloraz manganese.

» When inoculum levels are low, 0.2% dithiocarbamates (Zineb and Maneb) are also effective at preventing disease spread.

## Dry bubble, *Verticillium fungicola* var. *fungicola*

### Symptoms

» Primordial/pin head stage symptoms include a tiny, undifferentiated clump of tissue (up to 2.5 cm in diameter).

» At their mature stages, their formation is sometimes faulty, resulting in caps that are only partially distinct, stipes that are deformed into shaggy shapes, and caps that are positioned at an angle.

» Mushrooms with this condition are covered in a fine grey white mycelial growth; they are discoloured, dry, and do not decompose like mushrooms exposed to wet bubble conditions.

» The pileus, or cap surface, of a mature mushroom will have minute, pimple-like outgrowths (ranging in size from one centimetre to two) of a blue-grey colour.

» These blemishes typically have a bluish-grey or yellowish aura.

### Favorable conditions

» Diseases thrive at temperatures around 20 °C.

» At room temperature, the pathogen thrives.

### Survival and spread

» The majority of inoculum is introduced into a system through contaminated casing.

» Spores dispersed into the air by insects like flies, mites, and pickers can also act as a key source of inoculum.

» Another crucial method of propagation is watering.

## Management

» Pay close attention to sanitation in order to stop the spread of this disease.

» Be sure the dirt you use for the casings is clean and safe to use.

» Put the medium you'll be using to casing through a steam/formalin treatment to make it edible.

» The infected mushroom must be thrown out immediately.

» It's important to get rid of the flies and mites.

» Infected soil should be steamed or chemically treated to kill any remaining pathogens.

» The inverted glass containing the salt should extend out from the diseased area by at least 5 cms.

» Dry bubble disease can be effectively managed with the use of prochloraz manganese.

» To treat the casings, spray Zineb 0.2% three times.

## Green mold, *Trichoderma harzianum*

## Symptoms

» Bags, trays, and the undersides of shelves may show green sporulation patches.

» They grow into a dark green and become less noticeable.

» All of these are subsequently flushed down into the casing.

» Mushrooms that have broken through this hull are severely discoloured, misshapen, and perish quickly.

» As blank spots start appearing on the bed, production drops significantly.

## Favorable conditions

» A moist, dark environment rich in carbohydrates is ideal for fungi.

## Survival and spread

» Wind, spent compost dust particles, mites that have fed on *Trichoderma mycelium* in infected compost, hands, clothing, and machines are all vectors for the spread of Trichoderma spores to newly freshly heated compost.

» Infections can live on a wooden shelf they can colonize the usually damp wood.

## Management

» The cleanliness of the property must be prioritized above all else.

» The 0.05% benomyl spray has been shown to be effective.

» Use table salt as a localised antiseptic.

» At harvest's end, use steam or formalin to remove compost from beds.

» Formalin solution should be used on a regular basis to clean all tools and equipment.

» To enter the mushroom home, you need immerse your feet thoroughly.

» To stop the spread of disease, efficient mite control is essential.

## Olive green mold, *Chaetomium* spp.

### Symptoms

» Ten days after spawning, a fine greyish-white mycelium develops either within the compost or as an aerial growth on the surface.

» Most of the time, the spawning process starts out slowly and then slows even further.

» Late in the spawning season, fruiting structures resembling grayish-green cockle-burns, 1/16 inch in diameter, appear on straw in discrete areas of the impacted compost.

» The compost pile will stink like old shoes.

» The growth of *Chaetomium* spp. is typically encouraged in compost that does not promote spawn growth.

» Incubation times are longer than usual, delaying spawn development.

» Spawning bottles also show *C. globosum.*

### Spread

» Air currents, clothing, and other things used in mushroom cultivation can all disseminate ascospores.

### Management

» In order to achieve optimal results, compost needs to be adequately pasteurized and conditioned with a steady flow of fresh air.

» It has been suggested that Zineb (Dithane Z-78) 0.2% spray be used to get rid of olive-green mould.

» The yields of Agaricus bisporus were increased, and they were protected to a substantial degree from *Chaetomium olivaceum* thanks to the introduction of a thermophilic *Bacillus* sp.

## Bacterial blotch or brown blotch, *Pseudomonas tolaasi*

### Symptoms

» Appearance of dark brown, asymmetrical patches on the pileus's surface.

» These patches could be a lighter shade of brown at first, but they'll likely darken over time.

» Most of the time, the patches will have a satiny gloss and look slightly sunken.

» Mushrooms that have been severely damaged may become misshapen, and the entire crop may become a dark brown color and split in afflicted regions.

» It's possible that the stalk will rot away in the worst circumstances.

### Favorable conditions

» Extreme heat, excessive humidity, and constant cap-wetting.

### Survival and spread

» It's easy for pathogens to hang about in unsanitary environments such on trash, in tools, or in buildings.

» Flies, mites, and movable machinery in cropping rooms are the main vectors of transmission.

» As well as being spread by flies and pickers' hands, bacteria can hitch a ride on garbage, mushroom spores, water droplets, and even debris.

» When there is an infection in the beds, watering will spread the bacteria quickly.

### Management

» Before being used, the casing material must be treated with formalin or steam.

» In order to keep the caps from getting soggy and mouldy, watering should be done infrequently and not for too long.

» Beds are sprayed with bleaching powder (sodium hypochlorite) on a regular basis.

» The effective chlorine concentration that should be used in every watering is 150 ppm.

# References

Fletcher, J. T., & Gaze, R. H. (2007). *Mushroom pest and disease control: a color handbook*. Elsevier.

Grewal, P. S. (2007). Mushroom pests. In *Field Manual of Techniques in Invertebrate Pathology: Application and evaluation of pathogens for control of insects and other invertebrate pests* (pp. 457-461). Dordrecht: Springer Netherlands.

Keil, C. B. (1991). Field and laboratory evaluation of a Bacillus thuringiensis var. israelensis formulation for control of fly pests of mushrooms. *Journal of Economic Entomology, 84*(4), 1180-1188.

Lewandowski, M., Kozak, M., & Sznyk-Basałyga, A. (2012). Biology and morphometry of Megaselia halterata, an important insect pest of mushrooms. *Bull Insectol, 65*, 1-8.

Maurya, A. K., Murmu, R., John, V., Kesherwani, B., & Singh, M. (2019). Important Diseases and Pests of Mushroom. *Agriculture & Food: E-Newsletter, 1*(12), 189-193.

Richardson, P. N., & Grewal, P. S. (1993). Nematode pests of glasshouse crops and mushrooms. *Plant parasitic nematodes in temperate agriculture.*, 501-544.

Rinker, D. L. (2017). Insect, mite, and nematode pests of commercial mushroom production. *Edible and medicinal mushrooms: technology and applications*, 221-237.

Symes, C. B., & Chorley, J. K. (1921). Insect Pests of Mushrooms. *Fruit Grower, Fruiterer, Florist and Market Gardener, 51*(1320).

# 6

# Cabbage and Cauliflower

## INSECT PESTS

### Diamondback moth (DBM), *Plutella xylostella*

#### Damage symptoms

- » Cauliflower and cabbage are especially vulnerable to this pest from January to June and during dry monsoon spells.
- » To eat, newly born caterpillars drill holes through the underside of leaves.
- » The third and fourth instar larvae are the only ones that feed externally from the underside of leaves, protecting their dorsal surface with a transparent cuticular covering.
- » Leaf skeletonization occurs in the most extreme cases.

#### Pest identification

- » Eggs are placed singly on young leaves and are a pale golden colour.
- » The fully developed larvae are around 8 mm in length, pale yellowish green in coloration, and covered in fine hair.
- » When at rest, the wings of a diamond back moth fold together to form a white patch that resembles a diamond on the insect's back. The moths themselves are small and brownish-gray in colour.

#### Management

- » Cucumber, bean, pea, tomato, and melon as crop rotation.
- » Wet down the crop canopy with a spray of Neem seed powder extract 4%, Neem soap 1%, or Pongamia soap 1%.

» If adult population management is needed, light traps can be set up. Use 3–4 light traps for a plot size of 1 acre (60 or 100 W bulbs).

» Because this pest has become immune to many commercially available insecticides, it is imperative that you use the most recent and effective formulations.

» After planting 25 rows of cabbage or cauliflower, plant two rows of Indian mustard as a trap crop and spray it with 4% NSKE at the bud stage for integrated pest management. Two further 4% NSKE sprays at intervals of 10-14 days after the initial spray are permissible (Srinivasan and Krishna Moorthy, 1991).

» *Cotesia plutellae*, a parasitoid, has a control range of 70–80%, and *Didegma semiclausum*, another parasitoid, may provide outstanding parasitism of up to 68%.

» The 0.5–1.0 kg/ha range of *Bacillus thuringiensis* formulations (Dipel, Thuricide) has been shown to be effective.

» Besides *B. bassiana* and *P. farinosus*, other pathogens are also efficient.

## Leaf webber, *Crocidolomia binotalis*

### Damage symptoms

» The young larvae initially feed communally, but subsequently form a web around the leaves and feed inside.

» Cauliflower and cabbage heads spoil as a result.

### Pest identification

» The moth's forewings are a pale brownish color with noticeable wavy lines and wavy dots, while its hind wings are a semi-hyaline tone.

» Larvae are green with a crimson head and red stripes running lengthwise down their bodies. A length of 2 cms.

### Management

» The spread of the disease can be slowed by removing and destroying webbed bunches of leaves.

» Egg masses and gregarious larvae are gathered and disposed of.

» Neem seed powder extract (NSPE) or Neem seed kernel extract (NSKE) at 4% should be sprayed.

- » It is effective to dust the crop with Carbaryl (4% dust) or spray it with Malathion (0.05%).

- » Cartap hydrochloride 50 SP 500 ml, Spinosad 45 SC 125 ml/ha, or Azadirachtin 0.03% 2.5-5.0 L/ha. Phosalone 50 EC 1.0 L, Fenvalerate 20 EC, Cypermethrin 10 EC, Deltamethrin 28 EC, 250 ml. Avoid using insecticides that work in the same way twice.

- » For example, you may use Chlorfluazuron 5.4% EC at a rate of 0.1 ml per litre of water, or Indoxacarb 14.5% SC at a rate of 0.5 ml per litre of water, to spray the area.

- » The larval parasitoids *Microbracon mellus* and *Apanteles crocidolomiae* control the pest population.

## Head borer, *Hellula undalis*

## Damage symptoms

- » First, caterpillars eat away at the leaves' veins.

- » Afterwards, they emerge from the leaf's silky crevices to eat on its surface.

- » Cauliflower and cabbage heads are eaten as they expand.

- » Heavy infestations leave plants infected with worms and with malformed looking heads.

- » The mature larvae dig into stems, blocking the development of a single head, leading instead to the development of several.

- » In the months of March through July, the damage is typically at its worst.

## Pest identification

- » Pearly white when laid, the eggs will change pink the following day and then brown.

- » The larvae have a whitish-brown tint with 4-5 longitudinal stripes of a deeper purple.

- » For adults, it's a tawny greyish brown. The forewings are grey with wavy lines, a white tip, and a dark moon shape (lunular) at the edge, while the hindwings are pale dusky with a small fuscous suffusion at the tip.

## Management

- » In the early stages of an infestation, collecting and killing caterpillars mechanically can help.

» Neem seed powder extract (NSPE) 4% spray-on

» A 1% Malathion spray or 4% Carbaryl dust applied to the crop can effectively eliminate the larvae.

» The wasp, *Chelonus blackburni*, lays its eggs inside a cabbage webworm egg and raises its young inside the webworm caterpillar.

» Application of *Bacillus thuringiensis* at a concentration of 2 g/liter should be made during the earliest developmental stages.

## Aphids, *Brevicoryne brassicae*, *Lipaphis erysimi*

### Damage symptoms

» During the months of February through June, *Brevicoryne brassicae* is more common than *L. erysimi*.

» While they feed, nymphs and adults sucking cell sap and devitalizing plants, which negatively impacts the quality of the resulting head or curds.

» The affected areas turn a sickly colour and develop abnormalities.

» It is common to see aphids on the underside of leaves and at the plant's tips.

» When plants are severely infested, they might dry out and die.

» Damage from feeding causes the leaves of larger plants to curl and yellow, growth to be stunted, and the heads to become misshapen.

» The plant's roots will have a white cast skin.

### Pest identification

» Nymphs are pear-shaped and either winged or wingless and have a yellowish green coloration. They are very little (2.0–2.5 mm).

» In contrast to *L. erysimi*, *B. brassicae* is not as dark and has a white mealy.

### Favorable conditions

» Because high humidity encourages the rapid reproduction of this insect, conditions like these often occur together.

### Management

» We recommend planting the seeds as early as possible, ideally no later than the third week of October.

» Put out 30 yellow sticky traps per hectare to keep track of the aphid population.

» The recommended dosage for spraying Dimethoate or Phosalone is 2 ml per litre of water.

» As soon as aphids are spotted on the crop, spray it with Acetamiprid 20% SP at a rate of 0.2 g/liter of water.

» Neem seed powder extract (NSPE) 4% spray-on

» Mustard can be used as a trap crop in integrated pest management plans; after flowering, it attracts all aphids, protecting the main crop from damage.

» Protect parasitoids like the adult and juvenile *Aphidius colemani*, *Diaeretiella* and *Aphelinus* species, and others.

## Cabbage butterfly, *Pieris brassicae*

### Damage symptoms

» Caterpillars are the cause of damage.

» The leaves of the host plants are lacerated and skeletonized by newly hatched caterpillars.

» Adult caterpillars devour the host plant's leaves, starting at the leaf margin and working their way interior.

» Sometimes entire plants are consumed.

### Pest identification

» Larvae start out a pale yellow colour but mature into a greenish yellow color. The top of the head is black, and there are black patches all over the back. Short hairs all over the body make for an attractive appearance. Adult caterpillars reach a length of 40-50 mm.

» Male and female adults are both pure white, albeit males are often smaller. The wingspan of a mature female butterfly averages about 6.5 cms. Both sexes have a black patch at the leading edge of their forewings, but males also have black markings under their wings.

### Management

» Having fine-mesh netting in the nursery will prevent butterflies from reaching the crop and laying eggs on it.

» Mechanically collecting and killing caterpillars, typically found on the underside of leaves, by hand in the field.

» Reduce this pest population by spraying with Malathion (0.05%) or Diazinon

(0.02%) three weeks before harvest.

» The larva of the cabbage butterfly can be controlled biologically with the introduction of parasitic insects such the *Trichogramma*, *Apanteles glomeratus*, and *Pteromalus puparum*.

» Apply a 1-2 g/l solution of a commercial *Bacillus thuringiensis* formulation.

## Tobacco caterpillar, *Spodoptera litura*

### Damage symptoms

» The larvae eat the leaves and wreak havoc on the cauliflower and cabbage heads.

» Beet roots and radishes should have their leaves harmed.

» Very gregarious initially.

» Mature caterpillars are a dark grey or brown in colour.

» They do most of their damage at night and can be very destructive.

### Pest identification

» Egg globules have a yellowish-brown colour.

» Cattle moth larvae are pale green with a black head or patches.

» The brown adult wings have white, wavelike markings. The edges of the white rear wings feature a brown spot.

### Management

» Turn over the earth and plough it to reveal the pupae so they can be killed.

» Plant castor as a trap crop in border areas and along irrigation canals.

» In order to remove the hibernating larvae, one must flood the field.

» Light traps should be placed at a rate of 1 per hectare.

» To catch male moths, set out 15 pheromone traps (Pherodin SL) per hectare.

» Egg masses in castor and tomato should be collected and destroyed.

» Collect adult larvae by hand and dispose of them.

» With a concentration of $1.5 \times 10^{12}$ POBs per hectare, 2.5 kg of crude sugar, and 0.1 % Teepol, spray Sl NPV.

» To poison bait, combine 5 kg of rice bran, 500 grammes of molasses or brown sugar, 500 grammes of carbaryl 50 WP, and 3 litres of water per hectare.

Keep the bait in a container in the field overnight.

» You can apply Malathion 50 EC 1 liter/ha, Dichlorvos 76 WSC 1 liter/ha, or Chlorpyriphos 20 EC 2 liters/ha as a spray.

» Apply a 1-2 g/l solution of a commercial *Bacillus thuringiensis* formulation.

## Semilooper, *Trichoplusia ni*

### Damage symptoms

» Adults skeletonize the leaves by chewing big, uneven holes all over the plant, and larvae do most of their harm by feeding on the undersides of leaves.

» Cauliflower heads can become unsellable if contaminated with excessive levels of dark frass, and the same is true for cabbage and broccoli.

### Pest identification

» The adult form of the cabbage looper moth has a spotted brownish grey colour with a silvery figure eight pattern on the forewings and a tiny tuft of hair at the back of the head.

» The larvae of the looper are easily recognisable by their "looping" behaviour and distinctive pale green coloration with white stripes down each side.

### Management

» Eliminate caterpillars by hand.

» Adults can be lured into a light trap and killed this way.

» Use either 0.5% Quinalphos, 0.1% Endosulfan, or 0.1% Malathion spray.

# DISEASES

## Club root, *Plasmodiophora brassicae*

### Symptoms

» Yellowing of leaves, stunting, and eventual death are the visible symptoms of this disease.

» The young plants don't make it, and the older ones don't produce any good-looking heads.

» A club-shaped deformity is established on the roots and rootlets, which take the form of a spindle with a thick middle and tapered ends.

» The secondary, weaker parasitic organisms invade the club-shaped cells, disintegrating them to release toxins that ultimately destroy the plants.

## Favorable conditions

- » From 12 to 27°C, it takes place (25°C optimum).
- » Very wet ground.
- » Soil pH of 5.0–7.0 is neutral to acidic.

## Survival and spread

- » Primarily: dormant spores carried by the soil that can live for much longer there (10 yrs.) Broccoli, Brussels sprouts, cabbage, cauliflower, Chinese cabbage, mustard, radish, and turnip are all plants that might act as "collateral hosts."
- » Secondary: Irrigation water or root contact can spread dormant spores or zoospores.

## Management

- » When planting, we stay away from diseased fields.
- » Plants should be fungus and pest-free.
- » Disease is reduced when the pH is just little above neutral (about pH 7.2).
- » To raise the pH of the soil to 7.2 (6 weeks prior to planting, at a rate of 1.5 t/acre), hydrated lime should be added.
- » The use of too much water in irrigation should be avoided.
- » Rotate crops like potatoes, tomatoes, beans, and leeks over a lengthy period of time.
- » Seeds were treated with 4 g/kg of Captan or Thiram, then with 4 g/kg of Trichoderma viride.
- » Copper oxychloride soil drench at 0.25 %.

## Alternaria leaf spot, *Alternaria brassicicola*

## Symptoms

- » Dark spots up to 1 cm in diameter emerge on leaves.
- » Conidiophores form concentric rings on the lesion when the air is humid.
- » Petioles, stems, and seed pods are all covered in linear dots.
- » Brown spots develop on cauliflower curds.

» The dots on *A. brassicae* are less noticeable in size and have a paler color.

## Favorable conditions

» Around 26-28 °C in the soil.

» A pervasive dew or a high humidity level.

» Wet, cloudy, and occasionally rainy.

» Inoculation requires 9 hours of dew or rain.

## Survival and spread

» Mycelium that survives in the seed or spores that land on or are carried by the seed are the first line of defence against a fungal infection.

» Conidia spread by the wind or insects.

## Management

» Seeds are soaked in hot water (50°C) for 30 minutes.

» Keeping away from any overhead watering.

» Three-year agricultural cycles that do not include crucifer crops or cruciferous weeds like wild mustard.

» Use either Zineb 75% WP (600-800 g) or Mancozeb 75% WP (600-800 g) in a solution of 300-400 litres of water per acre.

## Downy mildew, *Peronospora parasitica*

## Symptoms

» On the undersides of the leaves, you may see a few little brown specks with a purple tinge.

» Downy growth on the underside of the leaves, and tiny, angular dots of pale yellow on the upper surface.

» The spots join together, and the leaves wither and fall off before their time.

» These areas allow soft rot to set in on cabbage heads.

» The curds of a cauliflower look brown.

» Blackening of the vascular system, dark brown and depressed lesions or streaks, and the eventual development of downy fungus growth are all signs of fungal infection in stems.

## Favorable conditions

- » It takes place at mild temperatures of 12-27⁰ C.

- » Very wet ground.

- » Soil pH of 5.0–7.0 is neutral to acidic.

## Survival and spread

- » The primary vector is oospores found in contaminated soil or decaying plant matter.

- » Sporangia that have been blown around or drenched by rain are secondary.

## Management

- » Once you've finished harvesting, pick up all the leftover debris.

- » Prevent heavy planting and constant moisture.

- » Switch out your brassicas for three years.

- » Don't water the lawn from above.

- » Open up the space so that air can flow freely (i.e. wide spacing, rows parallel to prevailing winds, not close to hedgerows).

- » Use of an effective fungal spray.

## Bottom rot, *Rhizoctonia solani*

## Symptoms

- » When transplanted cabbage plants grow large enough to cast shadows on the ground, the disease first emerges on the underside of the head leaves that are in contact with the soil.

- » As a leaf is being eaten, the midrib is usually the primary target.

- » Lesions that develop as a result are depressed, dark, and sharply elliptical, with the long side of the lesion running parallel to the patient's midrib.

- » As the weather gets drier, lesions may dry up and take on a papery brown look.

- » The mycelium may cover the lesion's surface in a sparse, weblike fashion.

- » Ultimately, a widespread black rot spreads from the leaf's base outward.

- » At first, only the leaf's tip turns yellow; eventually, the entire leaf shrivels.

- » When a plant is infected, its leaves fall off and it is left with just a stalk and a tiny little head.

> » When temperatures are warm and the relative humidity is high, bottom rot transforms into head rot.

> » The plant's stem is unharmed, thus the top stays in its normal position.

## Favorable conditions

> » Wet soil, damp or wet foliage, and temperatures between 20 and 28 °C are ideal for the growth of many plant diseases.

## Survival

> » Contaminated seed, weeds, and agricultural leftovers of cruciferous plants are also potential entry points for the fungus.

> » Three-year soil crop residue survival.

## Management

> » Incorporation of pesticide-free seed.

> » When the disease is particularly bad, farmers will often rotate out cruciferous crops for four years.

> » The plant bed should be placed in a sunny, well-drained area. Faulty drainage encourages the spread of disease in young plants.

> » Don't use too much nitrogen fertiliser. Infections are more likely to spread rapidly through succulent plants.

> » Crucifer seed should be sown as shallowly as possible when the soil temperature reaches 21 °C to ensure rapid germination and emergence.

> » If transplanted seedlings develop wire stem disease, they should be thrown away.

> » Gather mature cabbages as soon as possible. The longer cabbage leaves are left out in the field, the more damage they sustain.

## Cabbage yellows, *Fusarium oxysporum* f. sp. *conglutinans*

## Symptoms

> » Affected plant life becomes chlorotic and brown.

> » Uneven growth causes some leaves to seem misshapen.

> » Premature leaf death and senesce may occur, usually at the plant's foundation.

» Because of this pathogen's invasion, the vascular tissue of host plants will turn a dismal shade of brown or yellow.

» Those plants that survive often have stunted growth and lopsided leaf or stem yellowing.

» In time, the plant turns a sunny yellow.

» One side of the plant may show signs of withering earlier than the other.

## Favorable conditions

» Temperatures above normal seem to accelerate the spread of the disease.

» Disease spread is slowed when the temperature is below 20 °C.

## Survival

» This fungus can live in the soil and develop spores that can stay dormant for years.

## Management

» Incorporation of pesticide-free seed.

» Rotate non-host crops in every so often.

» Preserve the cleanliness of the playing field.

» This disease can be managed most efficiently by the use of resistant cultivars.

## *Cercospora* leaf spot, *Mycosphaerella (Cercospora) brassicicola*

## Symptoms

» Spots on leaves might be anywhere from a very light green to a completely white tint, and they're always surrounded by a brown tissue border and general chlorosis.

» The shape of the lesion, whether round or angular, might vary widely.

» Symptoms are most noticeable on the outer leaves.

» Spots on the leaves range in size from 1 centimetre to 2 cms in colour from brown to tan.

» Several greyish fruiting bodies emerge in concentric rings around the greyish core.

» There is a green ring around the spots that stays green long after the rest of the leaf has turned yellow.

» Defoliation may occur in severely afflicted plants.

» When the spot is on the head, the price drops.

**Favorable conditions**

» Temperatures and humidity levels between 13 and 18 °C are most conducive to the spread of disease.

**Survival and spread**

» The fungus can spread by seeds, but it usually just hangs out in weeds and uninvited plants.

» Wind, rain, irrigation water, and machinery can all have a role in the propagation of spores.

**Management**

» Eliminating and then collecting dead plants.

» The seeds were subjected to a 45 °C water bath for 20 minutes.

» Eliminate any weeds or volunteer cruciferous plants.

» When used early and frequently, fungicides can keep diseases under control.

**Stalk rot, *Sclerotinia sclerotiarum***

**Symptoms**

» Cabbage's stem and leaves develop wet areas low to the earth.

» In just a week to ten days, the entire plant will wilt and die.

» Cottony white fungal growth with numerous hard black sclerotia is characteristic of a head infection.

» Cauliflower leaves turn yellow from their tips all the way to their bases.

» Premature leaf fall is occurring.

» The stem rots, the stem girdles, and the stem rots up to the curd region.

» In addition, curds are impacted. The damaged areas have developed a white, fluffy mycelial growth that is full of sclerotia.

**Favorable conditions**

» The ideal conditions for the spread of disease are wet soil and temperatures between 10 and 25 °C.

**Survival**

> » This fungus may survive for a long time in the soil because to its sclerotia.

**Management**

> » After harvesting, collecting and burning all crop waste.
> » The infected plants must be uprooted and destroyed.
> » We are avoiding using sprinklers because of the inconvenience they cause.
> » Ensure proper waste management and extensive crop rotations away from host plants (Rice).
> » Sow seeds in order to encourage water to drain away from plant roots.
> » In order to eradicate sclerotia, warm-weather flooding must be applied to fields for extended periods of time.
> » Take care of weeds.
> » Use sprays of fungicides like Topsin, Ortiva 250 SC, Switch 625 WG, and Rovral 500 SC to get rid of this disease.

**White rust, *Albugo candida***

**Symptoms**

> » White, elevated pustules, caused by a fungus, appear under the plant's epidermis and are easily recognisable on infected leaves and flower parts.
> » These pustules, which look like blisters, might cause the stem, leaves, or flowers to develop in a twisted, abnormal manner.
> » When ready, the epidermis covering the pustule will break, releasing the sporangia, which can be spread to nearby host plants by wind or splashing water.
> » When leaves are severely diseased, they might wither and perish.

**Pathogen identification**

> » The conidiospores of the Albugo fungus are strung together.

**Management**

> » The process of collecting and burning crop waste after harvest.
> » Sow only wholesome seeds.
> » In order to lessen the impact of diseases, it may be helpful to keep leaves dry.

» Do not plant in areas where white rust has previously been a problem, as the oospores that cause the disease can travel through the soil.

» Use a spray containing Mefenoxam, Chlorothalonil, Fosetyl-Aluminum, Ridomil Gold MZ, or copper oxychloride.

## Black leg, *Phoma lingam*

### Symptoms

» Happens where there's rain during the growing season.

» Discolored patches are the first sign of a problem.

» The fungus' fruiting body, represented by tiny black dots, appears on the yellow patches.

» All of the plant's leaves may fall off and it may be destroyed in the event of a particularly vicious attack.

» When a vertical section of the damaged plant's stem is cut open, the sap clearly reveals a dark, almost black, discoloration.

» Rot spreads up through the entire root system.

» In the field, afflicted plants frequently topple over.

### Favorable conditions

» Wetter weather is ideal for it.

### Survival and spread

» As fungus can be spread from seed to plant, it is possible for it to infect young trees.

» Fungus can survive on crop waste and spreads via a number of different vectors (water, wind, animals, people).

### Management

» Establish a productive garden by sowing only wholesome seeds and seedlings.

» Infected seeds should be discarded.

» Boost ventilation and drainage in the soil.

» Copper oxychloride sprays on seed plants help protect against seed infections.

» Captan or Thiram + 4g/kg *Trichoderma viride* is a popular seed treatment.

» The process of collecting and burning crop waste after harvest.

» Dithane M-45 and Captan 80 WDG chemical processes.

» The Pusa Drumhead Cv. has been lauded for its resilience in the field.

## Powdery mildew, *Erysiphe cruciferarum*

### Symptoms

» Appears as powdery white dots, which can appear on either side of the leaf.

» Spots like these can quickly develop across a lot of leaves.

» Leaves turn a yellowish color, but no powder is produced.

» Infected leaves may progressively turn entirely yellow, die, and drop off due to powdery mildew.

» The plant's growth may be limited or its leaves may be defoliated, but it is rarely killed.

### Favorable conditions

» The crop's water stress, low relative humidity, and the presence of a thin film of moisture in which spores might germinate appear to be the most favourable conditions for this disease.

### Survival and spread

» Fungi are more likely to make it through the winter if they are able to cling to living plant material, which will allow them to develop new spores (conidia) in the spring.

» The wind can carry conidia over great distances.

### Management

» When feasible, choose a sunny spot to plant your garden.

» Guarantee adequate ventilation.

» Powdery mildew can be mitigated by overhead watering, as the spores are removed along with the water.

» Horticultural oils, Neem oil, Jojoba oil, Sulfur, and the biological fungicide Serenade are some of the least harmful fungicides on the market.

**Black rot, *Xanthomonas campestris pv. campestris***

## Symptoms

» Chlorotic to yellow lesions, formed like a 'V,' appear on leaf margins.

» They initially appear white but eventually turn black.

» Veins and veinlets darken to a deep blue.

» When the virus travels through the body, it eventually reaches the source.

» The stem's vascular bundle turns black, and then the fleshy stalk joins it.

» Deterioration is accelerated by interactions with soft rot organisms.

» Insects have infested and stained the cabbage and cauliflower.

## Favorable conditions

» Above 90% relative humidity.

» Very wet ground.

» Normal rainfall pattern being interrupted frequently.

## Survival and spread

» The black rot bacterium can hang around in the soil and on cruciferous weeds for up to two years if crop residue is left there.

» Internally propagated bacterial cells in seeds and the soil are the primary vectors of dissemination.

» Secondary transmission via floating bacterial cells carried by irrigation water and splashes of rain. It is possible for the bacterium to spread from weeds (pepper grass, wild radish, black mustard, wart cress, wild turnip) and neighbouring crucifer crops.

## Management

» If you want to avoid problems with *X. campestris* pv. campestris, make sure you're using clean seed.

» Crops other than crucifers should be rotated in every three years.

» Separation from commercial crucifer crops is essential for seed beds.

» Crops should be planted in soils with good drainage, and watering techniques should be implemented to reduce leaf wetness.

» The fields must be kept free of the cruciferous weeds.

» Steam or germicidal sprays should be used to disinfect seed beds and other equipment before planting.

» Reduce the insect population to slow the disease's progress.

» Agrimycin 200 ppm spraying.

# References

Ahuja, D. B., Ahuja, U. R., Srinivas, P., Singh, R. V., Malik, M., Sharma, P., & Bamawale, O. M. (2012). Development of farmer-led integrated management of major pests of cauliflower cultivated in rainy season in India. *Journal of Agricultural Science, 4*(2), 79.

Arora, R., Battu, G. S., & Bath, D. S. (2000). Management of insect pests of cauliflower with biopesticides. *Indian Journal of Ecology, 27*(2), 156-162.

Gruszecki, R., Walasek-Janusz, M., Caruso, G., Zawiślak, G., Golubkina, N., Tallarita, A., ... & Sękara, A. (2022). Cabbage in Polish folk and veterinary medicine. *South African Journal of Botany, 149*, 435-445.

Luther, G. C., Valenzuela, H. R., & Defrank, J. (1996). Impact of cruciferous trap crops on lepidopteran pests of cabbage in Hawaii. *Environmental Entomology, 25*(1), 39-47.

Larson, R. H., & Walker, J. C. (1939). A mosaic disease of cabbage. *Jour. Agr. Res, 59*(5), 367-392. McCulloch, L. (1911). *A spot disease of cauliflower* (No. 225). US Government Printing Office.

Moreb, N., Murphy, A., Jaiswal, S., & Jaiswal, A. K. (2020). Cabbage. *Nutritional Composition and Antioxidant Properties of Fruits and Vegetables*, 33-54.

Nagarkatti, S. U. D. H. A., & Jayanth, K. P. (1982). Population dynamics of major insect pests of cabbage and of their natural enemies in Bangalore District (India). In *Proceedings of the International Conference on Plant Protection in the Tropics. 1-4 March, 1982, Kuala Lumpur, Malaysia* (pp. 325-347). Malaysian Plant Protection Society.

Oates, M. J., Abu-Khalaf, N., Molina-Cabrera, C., Ruiz-Canales, A., Ramos, J., & Bahder, B. W. (2020). Detection of lethal bronzing disease in cabbage palms (Sabal palmetto) using a low-cost electronic nose. *Biosensors, 10*(11), 188.

Pajmon, A. (1999). Pests of cabbage. *Sodobno Kmetijstvo, 32*(11), 537-540.

Reddy, G. V. (2011). Comparative effect of integrated pest management and farmers' standard pest control practice for managing insect pests on cabbage (Brassica spp.). *Pest Management Science, 67*(8), 980-985.

Sharma, D., & Rao, D. V. (2012). A field study of pest of cauliflower cabbage and okra in some areas of Jaipur. *International Journal of Life Sciences Biotechnology and Pharma Research, 1*(2), 2250-3137.

Sharma, S. R., Singh, P. K., Chable, V., & Tripathi, S. K. (2004). A review of hybrid cauliflower development. *Journal of New Seeds, 6*(2-3), 151-193.

Sharma, S. R., Singh, P. K., Chable, V., & Tripathi, S. K. (2004). A review of hybrid cauliflower development. *Journal of New Seeds, 6*(2-3), 151-193.

Shiraishi, H., Enami, Y., & Okano, S. (2003). Folsomia hidakana (Collembola) prevents damping-off disease in cabbage and Chinese cabbage by Rhizoctonia solani. *Pedobiologia, 47*(1), 33-38. Subbarao, K. V., Hubbard, J. C., & Koike, S. T. (1999). Evaluation of broccoli residue incorporation into field soil for Verticillium wilt control in cauliflower. *Plant Disease, 83*(2), 124-129.

Srinivasan, K., & Moorthy, P. K. (1991). Indian mustard as a trap crop for management of major lepidopterous pests on cabbage. *International Journal of Pest Management, 37*(1), 26-32.

Tabone, E., Bardon, C., Desneux, N., & Wajnberg, E. (2010). Parasitism of different Trichogramma species and strains on Plutella xylostella L. on greenhouse cauliflower. *Journal of Pest Science, 83*(3), 251-256.

Tompkins, C. M. (1937). A transmissible mosaic disease of cauliflower. *J Agric Res, 55*, 33-46.

Tremblay, N., Bélec, C., Coulombe, J., & Godin, C. (2005). Evaluation of calcium cyanamide and liming for control of clubroot disease in cauliflower. *Crop protection, 24*(9), 798-803.

# 7

# Carrot

## INSECT PESTS

### Flea beetle, *Systena blanda*

### Damage symptoms

- » They create tiny pits in the leaves, giving the plant a "shot hole" appearance.
- » Seedlings and young plants are more vulnerable.
- » The adults of flea beetles cause the majority of the damage by feeding on the undersides of leaves, which results in the formation of small pits or irregularly shaped holes.
- » A high population density might be detrimental to young plants.

### Pest identification

- » Larvae have white bodies with brown capsules for heads, making them appear delicate and almost threadlike. Characteristically, they have big rear legs, which help them leap high distances.
- » Mature Chrysomelidae, which range in size from miniscule to modest, are distinguished by the striking enlargement of the femora of their hind legs.

### Survival

- » Hibernate as adults in wooded regions, windbreaks, and other places with plenty of leaf litter.

### Management

- » Flea beetles prey on young, defenceless plants, so it's best to get a head start and get your seeds in the ground as soon as possible.

» By enclosing the seed bed with floating row cover, you can prevent adults from accessing the area and laying eggs.

» Beetles can be kept from the soil by applying a thick layer of mulch.

» Diatomaceous earth and oils like Neem oil can be applied to the affected area to eliminate the problem.

» After harvesting, infested fields should have all weeds around the perimeter removed and the plant debris thoroughly disked.

» Use a pesticide such as esfenvalerate, carbaryl, spinosad, bifenthrin, permethrin, or diazinon.

## Crown and root aphids, *Dysaphis* spp.

### Damage symptoms

» Aphids like this like to cluster together at the base of the plant and at the top of the roots.

» Sometimes colonies will grow on the root just below the surface.

» Nevertheless, the greater risk is that their feeding could weaken the tops to the point that they would fall off during harvest, leaving the carrots in the ground.

### Pest identification

» The featherless creatures have a powdered wax coating and range in color from a light yellow to a bluish-green.

### Management

» Reducing populations of these aphids and the damage they cause requires careful sanitation and rotation to non-host crops.

» Imidacloprid, Cyfluthrin, or Malathion spray.

## Whitefly, *Trialeurodes vaporariorum*

### Damage symptoms

» Large numbers of silver leaf whitefly can be harmful to young plants.

» Honeydew, a byproduct of whitefly feeding, lends leaves a glossy, sticky appearance and makes them difficult to handle.

## Pest identification

» Crawlers are the first instars of nymphs, and they are nearly translucent, flat, and oval in shape. Its second nymphal instar (0.38 x 0.23 mm) is a see-through oval with undulating borders. The third nymphal instar is larger than the second (0.54 × 0.33 mm) but has a similar appearance.

» Adults' wings are initially transparent (0.75 x 1.10 mm), but they gradually develop a white wax coating. When compared to males, females are noticeably larger.

## Management

» Using a 1% Neem soap solution, spray the area.

» Some kinds of parasitic wasps, such as those in the genera Encarsia and Eretmocerus, are effective in reducing whitefly populations.

» Adult lady beetles, lacewing larvae, and big-eyed bugs are just a few of the predators that feed on whitefly nymphs.

## Rustfly, *Psila rosae*

## Damage symptoms

» The tunnels dug by the carrot rust fly, which are concentrated in the root's bottom two-thirds, are thinner and more tortuous than those dug by the carrot weevil.

» The larvae create holes in the roots, which eventually decay and spread disease.

» Other signs include drooping foliage and a change in colour from green to a rusty brown.

» Some young plants may die from the insect's intense attacks early in the growing season.

## Pest identification

» Maggot-like larvae can grow to a length of 10 mm and have a buttery golden colour.

» Here are a lot of similarities between puparium and seeds.

» The adult is a slim, glossy black fly that measures about 6 mm in length. It has a small but distinguishable reddish head, clear wings, yellow legs, and huge reddish brown eyes.

## Survival

» Larva hibernates in discarded root systems over the winter.

## Management

» Heavy ploughing in the summer.

» All crops in the carrot family should be rotated every three to five years (Apiaceae).

» Hosts in the wild are being wiped out.

» Installation of row coverings prior to the adult fly laying eggs on plants is essential for plant protection.

» Adult carrot rust flies can be tracked with orange and yellow sticky traps set in the field or in trees nearby.

» Carbofuran was spread at a rate of 1 kg active ingredient per hectare between 5 and 10 cms deep.

## Weevil, *Listronotus oregonensis*

## Damage symptoms

» Once the larva has matured and abandoned the carrot, the feeding tunnels will be found in the upper one-third of the root, with a prominent, darker, partly exposed tunnel in the crown.

» In order to enter the root system, young larvae either bore down from the stalk or crawl out of the earth.

» The tunnelling of developing larvae through the delicate roots of young plants can cause them to wilt and perish.

» Baby plants are a prime target for larvae.

» Root rot microorganisms can be introduced through feeding wounds.

## Pest identification

» The larvae look like reddish brown grubs with a creamy body.

» They reach a maximum length of around 6 mm as adults and have a dark brown, prominent snout with mottled colouring.

## Survival

» Adult weevils spend the winter in plant debris near affected carrot fields.

## Management

» Debris collection from Umbelliferous crops to lessen weevil breeding grounds.

» In other words, rotate your crops.

» If you want to grow carrots, don't sow them next to last year's harvest.

» At the second true leaf stage, the threshold for a single spray is between 1.5 and 5.0 weevils per trap.

» A second treatment at the four-leaf stage is required if the total number per trap is greater than five.

» The azadirachtin spray really works.

» Egg parasitism by *Anaphes* sp., a beneficial parasitoid, should be actively promoted as a means of pest management.

## Aster leaf hopper, *Macrosteles quadrilineatus*

## Damage symptoms

» Each instar drains plant vitality and produces white spots on leaves as it feeds on plant sap.

» When they feed, leafhoppers can spread the aster yellow fungus from plant to plant.

» Each time a leafhopper feeds, it increases the likelihood that another plant will get infected with aster yellows.

» Yellowing, stunting, profuse branching, and short internodes are all symptoms of infected plants.

## Pest identification

» The length of a nymph can be anything from 0.6- and 3.0-mm. Adults and juveniles alike share the same distinctive markings on their heads, however juveniles can be seen in a range of colours from yellow to light brown to a pale greenish-gray.

» The adult leaf hopper is typically 3.5–4 mm in length, but it can be as short as 2 mm or as long as 5 mm. This insect is also known as the six-spotted leaf hopper. The six black markings on the front of the skull of this species are in pairs.

## Survival

» Overwinter, likely as eggs and adults, on perennial weeds or fall-planted tiny grains.

## Management

» Floating row covers will keep them away from the carrot planting.

» Stick the yellow card out in the field when the plants are first sprouting in the spring.

» We need to spray the leaves starting the first week of July and doing so every 10 days until the end of August.

» Similarly, spray fields have boundaries.

# DISEASES

### Black root rot, *Thielaviopsis basicola*

### Symptoms

» Carrots cultivated in muck or highly organic soils are particularly susceptible to this disease.

» Carrots that have been washed and subsequently kept in plastic bags may develop black blemishes.

» Fungal growth, dark grey to black in colour, appears on the crown of the carrot, most noticeably at the remnants of the leaf bases.

» Roots may develop dark spots with ragged edges.

### Favorable conditions

» Carrot rot is typically a post-harvest issue that arises when carrots are not dried and cooled to less than 5°C before being packed.

» Soil spores are common, and they can multiply rapidly in cuts and bruises left by harvesting and packing.

» The development of diseases appears to be encouraged by root injury and warm storage temperatures.

### Management

» Practice extensive harvest cycles.

» Remains of harvested crops should be ploughed.

» Hydrothermal treatment of seeds.

## Cavity spot, *Pythium sulcatum, P. violae*

### Symptoms

» Carrots and parsnips are equally susceptible to cavity spot.

» The crop's total tonnage remains unaffected, but the roots are no longer saleable due to cavities.

» Carrots cultivated in either mineral soil or peat (muck) soil are susceptible to the disease.

» The tap roots of fully developed carrots show depressions that range in shape from elliptical to irregular.

» Root lesions are more common near the point where lateral roots branch from the main tap root.

» It is common for lesions, which start as little depressions, to grow larger as the roots develop.

» Carrots with sunken spots are most common in the final month before harvest, and their prevalence increases rapidly in over developed roots.

### Favorable conditions

» Caused by lack of drainage and acidic soils.

» Constant downpours and a cold, damp climate ideal for plant growth.

### Management

» Carrots being cultivated in a new location where the affliction has not yet been discovered.

» Use crops other than alfalfa and carrots in 3-year crop rotations.

» Be careful not to overwater.

» Carrots are more susceptible to infection the longer you wait to harvest them.

» Use Mefenoxam, Fenamidone, or Metam sodium to treat the soil.

## Cercospora leaf spot, *Cercospora carotae*

### Symptoms

» The symptoms first show up at the leaf margins, where they can be seen by observing the characteristic curling of the leaves.

» Inner leaf spots are tiny, generally round, and tan or grey to brown with a blank centre.

» The entire leaf shrivels and dies as the quantity and size of lesions grow.

» Younger leaves and plants are more susceptible to assault by fungi.

» Yet, in densely infested areas, even the youngest leaves are vulnerable to assault.

» Causes significant leaf and petiole blight in carrots.

» For severely diseased plants, entire leaves and petioles may wilt and die.

» Lesions on the petioles and stems are another sign of a pathogen's presence.

» It is possible for lesions to join together and girdle the stems, killing off the leaves.

## Favorable conditions

» Persistent rain and humidity during the growth season.

» With temperatures of about 28°C, growth is at its peak.

## Survival and spread

» Most spores hitchhike on seeds, although some can also be found in dead plants and on other hosts.

» Spread by wind and rain splash.

## Management

» Only plant verified disease-free seeds.

» In order to effectively control disease, seeds should be treated with hot water at 50°C for 15 minutes.

» Crop rotation, foliar fungicide applications, and ploughing under crop residues are all suggested methods of control.

» Rotating every two to three years is suggested.

» Reduce the spread of the disease by destroying diseased plant debris in the field.

» To prevent leaf blight, you can use a spray of either Foltaf (0.2%) or Copper oxychloride (0.3%).

## Leaf blight, *Alternaria radicina*

## Symptoms

» The irregularly shaped, dark brown to black spots first form around the leaflets' edges.

» Dark brown, coalescing lesions form on the petioles and stems, eventually encircling the plants.

» Whole leaflets may shrink and die as the disease advances, giving the impression that they were burnt.

» Lesions caused by the fungus Alternaria tend to appear first on older leaves and then on younger leaves.

» After the rows are closed, the disease spreads swiftly on the elder leaves of a crop that is developing.

» This is because the dense foliage traps moisture and prevents air from circulating among the older, lower leaves.

» Even tuber roots aren't safe from this disease.

## Favorable conditions

» Overhead watering or rainy weather, typically in the fall and winter.

» Despite the fact that 28 °C is the sweet spot for development, the infectious temperature range is much wider (about 13 to 35 °C).

## Survival and spread

» Infested agricultural leftovers serve as a fungus's breeding ground, and the spores can be spread by seeds.

» It can also live in the soil for at least 8 years without any crop residues.

» Microsclerotia are produced by fungi and act as a form of underground support.

» Typically, the disease is transmitted via its infective seeds in water.

## Management

» A soil with good drainage should be chosen.

» Prepare the seeds by soaking them in hot water (50 °C) for 15 minutes.

» Thiram seed treatment (3g/kg of seed).

» Non-host crops should be rotated in every three to four years.

» Elimination of diseased plants in the wild.

» Using a spray of Foltaf 0.2% or Copper oxychloride 0.3%.

» Disease prevention may be aided by furrow irrigation.

» Serenade ASO and MAX foliar sprays (*Bacillus subtilis* preparation).

## Powdery mildew, *Erysiphe cichoracearum*

### Symptoms

» Older leaves and petioles develop a white powdery growth that ultimately causes the leaves to discolour and wilt.

» It begins on older leaves and works its way to the younger ones until finally taking over the entire plant.

» The leaves become dark, curled, and brittle before they wilt and die under the intense disease pressure.

### Favorable conditions

» Summers are hot and dry.

### Survival and spread

» Wild carrots, celery, parsley, parsnip, and a variety of other closely related umbelliferous herbaceous plants serve as overwintering hosts.

» Conidia, which are released from spore-forming structures within powdery mycelium, aid in the subsequent development of the disease.

### Management

» At 4 weeks, spray fungicides containing 0.2% wetter sulphur; repeat at 7 and 10 weeks after sowing.

» Use a 0.1% Bavistin or 0.1% Benlate spray every 8-10 days.

» The carrot variety Arka Suraj is immune to powdery mildew.

## Sclerotinia rot/White mold, *Sclerotinia sclerotiorum*

### Symptoms

» As the disease has spread across millions of acres of canola and bean crops, there is now an abundance of the fungus' infectious ascospores.

» Carrots with the disease die off.

» The rate of damage to carrots in the field is modest, but after being washed and stored, white mould outbreaks are common.

» Even if just a tiny fraction of the roots are affected at first, the mycelium of the fungus can swiftly spread from one carrot to the next.

» Within a few weeks, the mould and sclerotia might cover every carrot in the container.

» The crown of infected carrots will show symptoms such as white mycelial growth and firm, black sclerotia (overwintering structures).

» During storage, the disease manifests as a soft, watery rot marked by the presence of fluffy white mycelia and black sclerotia.

## Favorable conditions

» Extreme fungal growth occurs in warehouses with humidity levels between 95% and 100%.

» Between 15 and 22 °C.

## Survival and spread

» Mold and fungi spread by their offspring.

» A fungal organism's ability to build sclerotia, or underground storage structures, ensures its long-term viability in the soil.

## Management

» The use of fungicides or hot water on seeds.

» Exchanging "host" crops for "non-host" crops every three to four years (grasses and onions are non-hosts).

» Keep an eye on it while it's in storage, and make sure it's kept at a cool, dry place with plenty of ventilation.

» Infected plants and tap roots should be ripped out of the ground and destroyed, both before and after being stored.

» Growing plants in containers.

» Springtime deluges after a winter of rain.

» A quick chill down before storing.

» Every parts of the storage facility must be meticulously cleaned.

**Bacterial leaf blight,** *Xanthomonas hortorum* **pv.** *carotae*

## Symptoms

> » Brown specks that are irregular in shape and typically appear first on the leaf 's edges.

> » In their early stages, lesions often have an uneven, yellow halo and a wet appearance.

> » Spots join together to form leaf blight, and dark brown stripes appear on the petioles of the affected leaves.

> » On the stems, water collects in elongated, squishy spots.

> » This blighting effect may spread to other areas of the flower.

> » It has been observed that some leaves and the petioles and flower stalks of some plants secrete a sticky, amber-colored bacterial fluid.

> » Infected plants may experience growth suppression.

## Favorable conditions

> » High rainfall and overhead irrigation have been linked to destructive outbreaks.

> » Disease flourishes between 25 and 30 °C.

## Survival and spread

> » The bacterium may live off of crop wastes in the soil.

> » While the bacterium can be found in seeds, it can also be found in irrigation water, runoff, on equipment, and in insects.

## Management

> » Xanthomonas-indexed seed should be planted.

> » To prepare the seed, soak it in boiling water at 52 °C for 25 minutes.

> » You can either furrow water or use a drip system. Don't water the lawn from above.

> » Rotate your crops every two to three years.

> » Formulations for spraying copper hydroxide.

**Soft rot,** *Pectobacterium carotovora*

**Symptoms**

» Bacterial infection causes the death of plants in the field.

» Tap roots develop soggy, asymmetrical sores.

» As the epidermis sometimes stays unharmed, soft lesions are sunken and a dull orange colour.

» Over time, contaminated root systems degenerate into a slimy, waterlogged mess.

» Most diseases appear after harvesting has already begun.

» Often accompanied with a strong stench.

**Favorable conditions**

» A long period of time at a high temperature (between 50 and 30 °C) is spent storing.

» There is a 90% chance of rain today.

**Survival and spread**

» Metaphorical hands-on transmission from plant to plant.

**Management**

» Disease can be effectively managed by spraying a solution of 2 grammes of carbendazim per litre of water every 8-10 days.

» Store items between 10 and 2 °C with relative humidity between 80 and 85 %.

**Common scab,** *Streptomyces scabies*

**Symptoms**

» Black, corky lesions that might be elevated or recessed, appearing horizontally on the root surface.

» Both lateral root growth and root perforation are potential sites for these.

» Surface carotene scabs develop on the root vegetable.

» The leaf symptoms are not obvious.

### Favorable conditions

» The superficial disease thrives in alkaline soils as well as dryish rich organic soils.

### Management

» The water supply should be managed.

» Avoid lime before sowing in scab-infested crops.

» It could be beneficial to rotate in non-root crops every so often.

» In other words, you shouldn't plant potatoes in fields.

» If the soil's pH is greater than 7, it needs to be lowered.

## Aster yellows, *Phytoplasma*

## Symptoms

» Younger leaves yellow, then turn crimson or purple.

» Abnormally short, gnarled stalks and a profusion of new shoots.

» Carrots with the disease are elongated and thin, and they have a dense, hairy growth of secondary roots.

» Roots of infected carrots are hairy, and their yellow, twisted leaves resemble ferns.

» Carrot flavour could be unpleasant as well.

## Spread

» *Macrosteles quadrilineatus*, a species of aster leaf hopper, has been identified as the primary vector.

## Management

» Get rid of the rouge plants before they spread.

» It is recommended to use the systemic pesticide Disyston 15G at the time of planting as a preventative measure.

» When leafhoppers are spotted in the field, use a foliar spray of carbaryl, diazinon, or pheromone.

# References

Uvah, I. I. I., & Coaker, T. H. (1984). Effect of mixed cropping on some insect pests of carrots and onions. *Entomologia experimentalis et applicata*, *36*(2), 159-167.

Davis, R. M., & Nu- ez, J. (2007). Integrated approaches for carrot pests and diseases management. In *General Concepts in Integrated Pest and Disease Management* (pp. 149-188). Dordrecht: Springer Netherlands.

Finch, S. (1993). Integrated pest management of the cabbage root fly and the carrot fly. *Crop protection*, *12*(6), 423-430.

Perron, J. P. (1971). Insect pests of carrots in organic soils of southwestern Quebec with special reference to the carrot weevil, Listronotus oregonensis (Coleoptera: Curculionidae). *The Canadian Entomologist*, *103*(10), 1441-1448.

Nunez, J., Hartz, T., Suslow, T., McGiffen, M., & Natwick, E. T. (2008). Carrot production in California.

Farrar, J. J., Pryor, B. M., & Davis, R. M. (2004). Alternaria diseases of carrot. *Plant disease*, *88*(8), 776-784.

Jayaraj, J., Wan, A., Rahman, M., & Punja, Z. K. (2008). Seaweed extract reduces foliar fungal diseases on carrot. *Crop Protection*, *27*(10), 1360-1366.

Chen, W. P., & Punja, Z. (2002). Transgenic herbicide-and disease-tolerant carrot (Daucus carota L.) plants obtained through Agrobacterium-mediated transformation. *Plant Cell Reports*, *20*, 929-935.

El-Tarabily, K. A., Hardy, G. E. S. J., Sivasithamparam, K., Hussein, A. M., & Kurtböke, D. I. (1997). The potential for the biological control of cavity-spot disease of carrots, caused by Pythium coloratum, by streptomycete and non-streptomycete actinomycetes. *The New Phytologist*, *137*(3), 495-507.

Dugdale, Mortimer, Isaac, & Collin. (2000). Disease response of carrot and carrot somaclones to Alternaria dauci. *Plant pathology*, *49*(1), 57-67.

# 8

# Celery

## INSECT PESTS

### Green peach aphid, *Myzus persicae*

#### Damage symptoms

- » Little, mushy-bodied insects that live on the undersides of leaves and/or stems.
- » Depending on the species and the plant they feed on, aphids can range in colour from green or yellow to pink, brown, red, or even black.
- » Yellowing, distortion, necrotic patches on leaves, and slowed growth in new shoots are all symptoms of a severe aphid infestation.
- » Aphids excrete a sugary fluid called honeydew, which promotes the development of sooty mould on the plants it infests. A decrease in crop value may result from contamination.
- » Aphids remove a lot of sap from plants, which stunts their development and decreases production.

#### Pest identification

- » Aphids can be identified by the cornicles (tubular structures) that extend posteriorly from their bodies.

#### Management

- » Infestations of aphids can be managed through selective pruning if they are confined to a small number of leaves or shoots.
- » Before planting, make sure the transplants are free of aphids.

» Tolerant cultivars should be used if they are available.

» Aphids can be discouraged from feeding on plants by using reflective mulches, such as silver coloured plastic.

» Spraying a forceful jet of water onto sturdy plants helps dislodge aphids from leaves.

» Chemical applications included 0.2 litres of Decis Mega 50 EC, 0.6 litres of Confidor Energy, 1.5 grammes of Mospilan 20 SG per 12 litres of water, and 25 litres of Actara WG.

» The most effective way of control is the use of insecticidal soaps or oils like Neem or Canola oil.

» Natural aphid control is provided by several species of parasitic wasps from the genera *Diaeretiella* and *Lysiphlebus*.

» Aphids are prey for lady beetles, syrphid flies, and lacewings, among other predators.

» After harvesting, destroy any crop remains promptly.

## Armyworm, *Mythimna unipuncta*

### Damage symptoms

» Result in single or clusters of holes of varying sizes and shapes in the plant material.

» Feeding in the crown and chewing away at the midrib and growth point are two ways in which celery seedlings are damaged.

» Young larvae will skeletonize leaves and cause shallow, dry lesions on fruit if they are fed on heavily.

### Pest identification

» It is possible to see egg clusters on the leaves, each containing anywhere from 50 to 150 eggs. These egg clusters have a cottony or fuzzy appearance because they are covered in a whitish scale.

» Older larvae are typically darker green with a dark and light line running along either side of their body and a pink or yellow underside, but younger larvae are generally pale green to yellow in appearance.

» A single white mark may be seen in the middle of the moth's buff-colored forewing, and its wingspan measures between 35 and 45 mm in maturity.

## Management

- » Kill larvae and pupae by discing fields right after harvest.
- » Burn or otherwise remove boundary weeds from fields.
- » Whilst chemicals exist for commercial management, many of those designed for use in the home garden are ineffective against the larvae.
- » Parasitoid wasps like *Hyposoter exiguae* and *Chelonus insularis*, as well as tachinid flies like *Lespesia archippivora*, are common.
- » Insects (e.g., minute pirates, big-eyed damsels, and assassin bugs), spiders (Araneae: Lycosidae and Phalangiidae), and beetles (e.g., ground beetles) are all potential predators (Carabidae).
- » *Bacillus thuringiensis* is used in this application.

## Leaf miner, *Liriomyza trifolii*

## Damage symptoms

- » If you look closely at the leaves, you can see mining galleries made by the larva as it eats them.
- » The mature insects consume the flower's stem.
- » Phytopathogenic substances could be implanted on the insect-caused lesions.
- » Damage from leaf miners in celery can cause premature senescence of the outer petioles, a later harvest, and a lower yield.
- » The aesthetic value of celery leaves declines after being mined by leaf miners.

## Pest identification

- » Less than 0.1 inches in length, the adult fly has yellow legs and clear wings. These are crimson eyes on a yellow head. As for the rest of the body, it's primarily dark grey and black.

## Management

- » Immediately following harvest, the field should be disked to bury and eliminate any infected crop residue.
- » Using Confidor Energy, Mospilan, Actara, Laser 240 SC, and Decis Mega EW 50 for treatment.

# DISEASES

**Early blight,** *Cercospora apii*

## Symptoms

» Unbounded brownish-gray patches that start off as small yellow flecks on the upper and below leaf surfaces.

» Little, brown, round dots on the leaves.

» The texture of lesions becomes papery.

» A dry, brown area can quickly grow to be at least 1 centimetre in diameter.

» A leaf that has been spotted frequently will eventually dry up, wither, and fall off.

» Petioles of celery plants are susceptible to infection by Cercospora under heavy disease burden.

## Favorable conditions

» Conditions favourable to the emergence of a disease include warm temperatures, high humidity, and persistent leaf wetness.

## Survival and spread

» Seeds can harbour the pathogen *Cercospora apii*, which has been found to live on celery plant remains in the wild.

» Water droplets and wind carry the spores to new locations. Moreover, celeriac is also a host for this virus.

## Management

» Reducing disease inoculum through deep ploughing of crop waste.

» Seeds that have been tested to be free of the fungus Cercospora should be used.

» Rotating non-host crops every three or four years.

» It may be advantageous to increase air circulation in the field by increasing the distance between rows, decreasing the number of plants per acre, or choosing cultivars with upright growth patterns.

» Fungicides should be used as needed (Propiconazole, Azoxystrobin, Chlorothalonil, Copper Hydroxide).

## Late blight, *Septoria apiicola*

### Symptoms

- » In older outer leaves and petioles, irregularly shaped chlorotic leaf spots begin as little, reddish-brown dots that eventually turn necrotic.
- » Spots of this type swiftly progress from brown to black, and it is not uncommon for multiple smaller lesions to converge into a single larger lesion.
- » A disease that causes leaf discoloration and eventually causes leaf death.
- » Embedded within lesions like tiny specks of pepper or mucky mud, tiny black flask- shaped creatures called pycnidia.

### Survival and spread

- » It is primarily spread through the consumption of infected celery seed.
- » Seeds can harbour fungus for up to two years if properly stored.
- » Rainfall, tainted irrigation water, and the transportation of people, objects, and animals are all potential vectors for the dissemination of fungi.

### Management

- » Try to find seed that hasn't been contaminated by disease.
- » Use only celery grafts guaranteed to be disease-free.
- » Seeds that have been treated with hot water.
- » Rooting out rogue plants that are carrying an infection.
- » Use of fungicides in the seedbed results in greater productivity and lower costs.
- » Crop rotations of three or four years, with the crop waste ploughed under deeply, to maximize fertility.
- » Increasing the amount of space between plants, decreasing the number of plants per area, or choosing varieties with upright growth patterns are all methods of increasing airflow within a field.

## Fusarium yellows, *Fusarium oxysporum* f. sp. *apii*

### Symptoms

- » The outer leaves on a celery plant will turn yellow if the plant's roots become infected.

» Eventually, the disease will cause the foliage to turn brown and die.

» Plant death or severe stunting characterized by yellowing.

» Brown, soggy roots.

» The stems' vascular tissue has turned a strange tint.

» Diseased plants quickly wilt.

» Infested soils hinder or destroy celery's growth.

## Favorable conditions

» Due to higher soil and air temperatures, summer celery crops are particularly vulnerable.

» If the soil pH is near to 7, diseases are more likely to occur.

## Survival and spread

» Once fungus is introduced into soil, it can live there continuously, typically through infected transplants or contaminated equipment.

## Management

» Maintaining a routine of sterilising tools and machinery.

» While planting, make sure to use seedlings that have not been exposed to any diseases.

» Putting seeds on soil that hasn't been treated for pests.

» The most effective method of pest management in affected fields is planting resistant or tolerant cultivars, such as Matador, Peto 285, Picador, or Starlet.

» To lessen spore populations in diseased fields, try rotating in onions or lettuce every two to three years.

## Bacterial leaf spot, *Pseudomonas syringae* pv. *apii*

## Symptoms

» Necrotic spots on leaf blades, often less than a quarter of an inch in diameter, are bounded by circular or angular veins and may be surrounded by a chlorotic halo.

» Faded and rust-colored lesions develop.

» Petiole discolouration and petiole streaking in brown are two indications of brown stem.

» Spots on leaves can be a variety of colours, from tan to brown to black.

» These bacterial leaf spots are systemic, meaning they may be seen from both the upper and lower surfaces of the affected tissue.

## Favorable conditions

» Conditions that are warm and damp are conducive to the growth of many diseases.

» When temperatures rise above 30 °C, bacteria flourish and spread, triggering severe disease symptoms.

## Survival and spread

» This dangerous pathogen can be found in the decomposing celery.

» A bacteria called *Pseudomonas syringae* pv. *apii* is spread through seeds.

» Pathogens can be introduced into crop fields via infected transplants.

» It only takes a sprinkle of water for a pathogen to spread quickly.

» Spread of bacteria between plants is facilitated by a lack of space between them.

## Management

» To avoid disease, it is important to plant indexed seed.

» A 25-minute soak in hot water will considerably reduce the number of infectious spores carried by the seeds.

» On the field, sprinkler irrigation should be avoided.

» In order to drastically lessen the spread of this disease, it is recommended to use seed that is at least two years old.

» Do use the area with copper fungicides.

## Soft rot, *Erwinia carotovora*

## Symptoms

» Brown, squishy, and depressed lesions develop towards the base of the petioles.

» The celery stem is the primary target.

» Brings about mushy deterioration and drenched celery leaves and stems, as well as an unpleasant stench.

## Favorable conditions

» Soil must be waterlogged for extended periods for diseases to emerge.

## Survival and spread

» Oxygen deficient plant tissue is a fertile breeding ground for bacteria.

## Management

» Let crop wastes to decay for an adequate amount of time before planting a subsequent crop.

» Celery needs soils with good drainage.

» Plants need to dry out between waterings.

» Use drip irrigation on the field.

» Keep harvesting cuts to a minimum to reduce post-harvest disease outbreaks.

» Planting celery next to cereals or other resistant crops is a good idea.

» The foliage of your plants has to be able to dry out quickly, so make sure you leave enough space between them.

## References

Huang, J. F., & Apan, A. (2006). Detection of Sclerotinia rot disease on celery using hyperspectral data and partial least squares regression. *Journal of Spatial science*, *51*(2), 129-142.

Jagger, I. C. (1921). Bacterial leafspot disease of celery. *Journal of Agricultural Research*, *21*(3), 185.

Jones, D., & Granett, J. (1982). Feeding site preferences of seven lepidopteran pests of celery. *Journal of Economic Entomology*, *75*(3), 449-453.

Lacy, M. L., Berger, R. D., Gilbertson, R. L., & Little, E. L. (1996). Current challenges in controlling diseases of celery. *Plant Disease*, *80*(10), 1084-1091. Meade, T., & Hare, D. J. (1995). Integration of host plant resistance and Bacillus thuringiensis insecticides in the management of lepidopterous pests of celery. *Journal of economic entomology*, *88*(6), 1787-1794.

Paula Wilkie, J., & Dye, D. W. (1974). Pseudomonas cichorii causing tomato and celery diseases in New Zealand. *New Zealand Journal of Agricultural Research*, *17*(2), 123-130.

Raid, R. N. (2004). Celery diseases and their management. In *Diseases of Fruits and Vegetables Volume I: Diagnosis and Management* (pp. 441-453). Dordrecht:

Springer Netherlands.

Reitz, S. R., Kund, G. S., Carson, W. G., Phillips, P. A., & Trumble, J. T. (1999). Economics of reducing insecticide use on celery through low-input pest management strategies. *Agriculture, ecosystems & environment, 73*(3), 185-197.

Weintraub, P. G., Arazi, Y., & Horowitz, A. R. (1996). Management of insect pests in celery and potato crops by pneumatic removal. *Crop protection, 15*(8), 763-769.

Van Steenwyk, R. A., & Toscano, N. C. (1981). Relationship between lepidopterous larval density and damage in celery and celery plant growth analysis. *Journal of Economic Entomology, 74*(3), 287-290.

# 9 Chilli/ Bell Pepper

## INSECT PESTS

### Thrips, *Scirtothrips dorsalis*

#### Damage symptoms

- » Symptoms looks like speckles on the leaves are actually thrips that have hatched from their eggs in the plant tissue.
- » Malformed, irregularly expanding leaves.
- » Spots or streaks of bleaching or browning on fruit.
- » Flower petals with silvery streaks.
- » Leaf discoloration: brown spots.
- » Adult thrips congregated in clusters along veins on the lower leaf surfaces.
- » The crinkled, curled leaves are a telltale sign of leaf curl virus.
- » A petiole that's a little too lengthy.
- » Fragile flowers fall to the ground.
- » When plants are infested at a young age, they are unable to fully develop and produce flowers or fruits.

#### Pest identification

- » A nymph's abdomen is small, linear, and frail, and it is a pale straw yellow.
- » In the adult stage, the wings develop a fringe.

**Management**

» To control the thrips population, intercrop with agathi (Sesbania grandiflora).

» Peppers shouldn't be planted right after sorghum.

» Do not plant a crop of chilli peppers and onions together.

» To prevent thrips from multiplying too rapidly, sprinkle water over the plants.

» This dark, silvery plastic can be used as mulch to help prevent thrips infestation.

» Imidacloprid 70% WS @ 12 g/kg of seed is the recommended treatment.

» The Carbofuran 3G are 33 kg/ha or Phorate 10G are 10 kg/ha.

**Aphid,** *Myzus persicae*

**Damage symptoms**

» Insects with soft bodies that live on the undersides of leaves and/or stems, typically in green or yellow.

» A strong aphid infestation can cause slowed growth, yellowing of the leaves, and necrotic patches on the leaves and stems.

» Honeydew, an adhesive sweet fluid secreted by aphids, promotes the development of sooty mould on plants.

» Leaves have crinkled and coiled up.

» As a result of the infestation, the affected plants become whitish and sticky looking.

**Pest identification**

» Green at birth, nymphs eventually develop a bright yellow.

» Adults are a drab yellowish green in color.

**Management**

» Aphids can be deterred from plants by using reflective mulches, such as silver coloured plastic.

» The most effective way of control is the use of insecticidal soaps or oils, such as Neem or Canola oils.

» Seeds should be treated with Imidacloprid 70% WS at 12 g/kg of seed.

» Use 10 kg per hectare of phorate 10% G.

## Pod borer, *Helicoverpa armigera*

### Damage symptoms

» An early instar larvae diet consists primarily of plant matter.

» Mature larvae are the primary fruit-borers.

### Pest identification

» The eggs are round, white, and placed individually.

» The larvae are a range of colours from green to brown.

» Pupae are brown in colour and can be found in places including dirt, leaves, pods, and crop waste.

» The adult female is a brownish yellow stout moth, while the male is a pale green with distinctive "V"-shaped markings.

### Management

» The diseased fruit must be gathered and destroyed along with any mature larvae.

» Put up 15 hectares of *Helilure pheromone* traps.

» The flowering period coincides with six releases of *Trichogramma chilonis* at a rate of 50,000/ha each week.

» *Chrysoperla carnea* should be released at a rate of 50,000 eggs or grubs / hectare, once a week starting at 30 DAS.

» To get rid of larvae, spray a mixture of cotton seed oil (300 g/ha) and HaNPV (1.5 x $10^{12}$ POB/ha).

» Carbaryl 50 WP Spray 2% g/L or 2% *B. thuringiensis*.

» Put out poison bait consisting of 1.25 kg of carbaryl, 12.5 kg of rice bran, 1.25 kg of jaggery, and 7.5 litres of water per hectare.

» The EPA recommends spraying 4 g/10 L of emamectin benzoate 5% SG, 6.0 g/10 L of flubendiamide 20 WDG, 6.5 ml/10 L of indoxacarb 14.5% SC, 7.5 ml/10 L of novaluron 10% EC, 3.2 ml/10 L of spinosad 45% SC, or 2.0 g/L of thiodicarb 75% WP.

» The larval population of *H. armigera* on chilli was effectively managed by planting one row of African marigold after every 18 rows of chilli and spraying with HaNPV (Shivaramu and Kulkarni, 2001).

### Gall midge, *Asphondylia capparis*

#### Damage symptoms

» Flower buds, flowers, and young fruits are all fair game for maggots.

» The damaged bloom withers and falls off its stem, but the infected bud never opens.

» The flower and fruit harvest has been severely reduced.

» Damaged fruit fails to mature.

» Less crop yield, smaller fruit, fewer seeds, and more deformed fruit.

#### Pest identification

» Adults are a pale, mosquito-like brown.

» Maggots in their mature state are pale yellow in colour, very short (only 3 mm in length), pointy on both ends, and limbless.

#### Management

» Insecticides containing neem oil should be applied at a concentration of 3 ml per litre.

» Neem cake (250 kg/ha) and Vermicompost (1 t/ha) applied to the crop before planting were as effective as the prescribed pesticides at reducing pest populations.

» An application of 250 grammes of Acephate 75 SP and 250 ml of DDVP 76 EC per litre of water.

» Both *Bracon* sp. and *Eurytoma* sp.

### Tobacco caterpillar, *Spodoptera litura*

#### Damage symptoms

» Newly hatched larvae, often known as neonates, tend to congregate in groups.

» They only remove the leaf's green inside and leave the epidermis intact.

» In order to eat, the second and third instar larvae bore holes in the leaves.

» Pods and leaves that have been skeletonized either fall off or turn a white tint when dried.

» The larvae feed on the scraped flesh of ripe fruits.

**Pest identification**

> » The adult moths of this species are a pale brown colour with white, wavy lines on the forewings and a brown patch on the white, hind wing edge.

> » The larvae look like little, worm-like grubs that are either grey or dark brown.

> » The pupae of the butterfly are a deep chocolate brown.

**Management**

> » For the purpose of destroying pupae, ploughing the soil is an effective method.

> » Gather the egg masses, eliminate the gregarious larvae and adult caterpillars.

> » Put up pheromone traps at a rate of 15 per hectare (sporelure).

> » Install sex pheromone traps Pherodin SL at a density of 12 units per hectare (ha) to track insect activity and time pesticide applications accordingly.

> » Castor is a trap crop that can be grown along borders and irrigation bunds.

> » Poison bait pellets made from 12.5 kg of rice bran, 1.25 kg of jaggery, 1.2 kg of carbaryl 50% WP, and 7.5 litres of water are dispersed around the fields in the evening, killing any caterpillars that emerge to feed.

> » It is suggested to use a spray formula of Chlorpyriphos 2.5 ml/liter of water, Quinalphos 2 ml/liter of water, Endosulphan 2 ml/liter of water, Thiodicarb 1 g/ liter of water, or Acephate 1 g/liter of water.

> » *Bacillus thuringiensis* at a concentration of 2 g/ lit.

> » In the evening, spray SlNPV at a rate of 1.5 x $10^{12}$ POB/ha.

**Yellow mite, *Polyphagotarsonemus latus***

**Damage symptoms**

> » Leaves wrinkle and curl as they fall.

> » Long-petioled leaves.

> » Growth arrest.

> » Fruits are also susceptible to the pest's infestation.

**Pest identification**

> » Eggs, of a white tint and an oval form.

> » Nymphs are a pure white in colour.

» Adults are often large, wide, and golden in colour.

**Management**

» The predatory mite *Amblyseius ovalis* should be encouraged to do its thing.

» Use 10 kg of phorate per hectare.

» We recommend spraying with either Dimethoate 30% EC at 1.0 ml/L, Ethion 50% EC at 2.0 ml/L, Oxydemeton -Methyl 25% EC at 2.0 ml/L, Phosalone 35% EC at 1.3 ml/L, or Quinalphos 25% EC at 1.5 ml/L.

# DISEASES

### Damping off, *Pythium aphanidermatum*

**Symptoms**

» Reduced seed germination and a weak stand of seedlings due to a disease that ravaged the nursery beds. Serious problems in germinating seeds, with death rates of 25–75%.

» Low levels of seed germination occur when seedlings die before

» The onset of a disease after seedlings have emerged from the soil but before their stems have lignified is known as post-emergence damping off.

» Infection forms a lesion at the collarbone.

» Browning and decay characterise infected areas.

» Plants wither and die because their tissues become mushy.

**Favorable conditions**

» High soil moisture and a temperature of 25 to 30 °C are the results of heavy rainfall, excessive and frequent irrigation, poorly drained soil, and close spacing.

» Rhizoctonia thrives in damp soil between 30 and 35 °C.

**Survival and spread**

» Primary: oospores in the soil for *Pythium* and sclerotia in the soil for *Rhizoctonia*.

» Secondary: *Pythium zoospores* in irrigated water. *Rhizoctonia* spreads its mycelium and sclerotia via irrigated soil.

**Management**

» To avoid bringing in fungal problems from seed, only use high-quality seed stock.

» According to their moniker, water moulds thrive in wet environments.

» Maintain a moist but not drenched soil, and pull up any symptomatic 3s.

» The seedlings need plenty of fresh air.

» To stop the condition from spreading further, use a gentle fungicide like copper- based product or even Chamomile Tea.

## Leaf blight, *Cercospora capsici*

## Symptoms

» *Cercospora* leaf blight manifests itself as a ring of dark brown around the periphery of a spot that is a pale grey or white colour.

» Defoliation can occur when numerous tiny areas clump together.

» Yellowing and defoliation of the leaves accompany a whitened area in the middle.

» The spot's centre can sometimes fade away.

» Stems and twigs also develop dark brown, irregular lesions that are surrounded by white.

» In extreme circumstances, twigs will die back.

» The fungus was found to be sporulating on the spots when given a damp environment.

» Infected plants experience significant defoliation and have trouble blossoming and fruiting as a result of the disease.

## Survival and spread

» Mostly, dormant mycelium in contaminated crop residues, seed stock, and stragglers.

» To a lesser extent, conidia were carried by the wind and distributed.

## Management

» Selecting seed from healthy plants is crucial for preventing first infection.

» After the fourth harvest of green chilies, a 0.2% Tricel spray can be used to keep the plant healthy.

» Protecting crops from disease requires a foliar spray of 0.2% Chlorothalanil.

**Powdery mildew, *Leveillula taurica***

**Symptoms**

> » A yellow powdery substance is seen on the underside of affected leaves, while the upper side turns a bright yellow.

> » When the disease progresses, a white mass might appear on both sides of the leaves.

> » Curling of diseased leaves, diminished size of infected fruit, and early defoliation are all symptoms of this ailment.

**Favorable conditions**

> » Both wet and dry environments are conducive to the spread of disease.

> » Conidial germination is facilitated by cool, dry conditions.

> » High RH encourages the growth of pathogens.

> » Pestilence moves quickly in damp environments, and it tends to attack older leaves.

**Survival and spread**

> » The active component is the dormant mycelium in the contaminated crop waste.

> » To a lesser extent, conidia were carried by the wind and distributed.

**Management**

> » It has been claimed that the dosages of 0.2% Baycor, 0.3% Devisulfur, and 0.05% Bayleton are all quite effective.

> » Penconazole (0.1%), Propiconazole (0.1%), Triadimefon (0.1%), Difenconazole (0.05%), and Hexaconazole (0.1%) were the next most effective at controlling PM and producing the highest yields of fruit, followed by Myclobutanil (0.05%) sprayed at the onset of disease and then twice more at 7-day intervals.

**Fusarium wilt, *Fusarium oxysporum* f. sp. *capsici***

**Symptoms**

> » Withering of the plant and the rolling of the leaves upwards and inwards.

> » When leaves turn yellow, they eventually die.

> » There is a temporary yellowing and wilting of the upper leaves followed by a permanent wilt in a few of days.

> » Lower stem and root vascular browning.

## Management

- » Incorporation of wilt-proof plant strains.

- » The use of a 1% Bordeaux mixture, 0.25% Fytolan, 0.25% blue copper, or 0.25% fytolan drench may provide some defence.

- » A formulation of *Trichoderma* viride weighing 4 grammes per kg of seed or 2 gm of Carbendazim per kg of seed are both effective treatments.

## Root rot, *Rhizoctonia solani*

### Symptoms

- » Seedlings are particularly vulnerable to Rhizoctonia root rot brought on by *Rhizoctonia solani*.

- » But, *R. solani* can also infect fully grown plants, causing root rot and ultimately death for chilli plants.

### Survival

- » Soil and decaying organic matter are ideal habitats for the fungus *Rhizoctonia solani*.

### Management

- » The use of fungicides like Thiram or Captan on high-quality seed and preventing soil saturation are essential for controlling this fungal infection conditions.

- » Soil fumigants and solarization are two methods that can be used to combat a large population of pathogens in a given field.

## Phytophthora blight, *Phytophthora capsici*

### Symptoms

- » The fungus spread across the entire plant.

- » Too much water in the soil, from either too much irrigation or too much rain, is a typical cause of problems.

- » The most noticeable indication of disease is wilting, which occurs when the root and lower stem become infected.

- » The pattern of wilting plants in furrow-irrigated fields is delineated by the rows.

- » Stem and collar rot, characterised by rapid wilting and the absence of foliar yellowing, is the most typical sign.

» With time, the entire field begins to show signs of wilting.

## Favorable conditions

» The spores of this disease are carried by the splash of water and can infect a whole row of plants.

## Spread

» In order to move from plant to plant, *Phytophthora capsici* creates cells that are designed to swim, and we call them zoospores.

## Management

» Growing plants from disease-free seeds.

» Grow plants from sterile cuttings.

» The use of non-host crops in a three-year crop rotation.

» Water shouldn't sit in the field for more than 12 hours after planting, and too much water can kill your seedlings if you water them in before they emerge.

» The gathering and eventual disposal of spoiled fruit.

» Moreover, 12 and 20 days after sowing, soaking the soil with a 1% Bordeaux mixture or 3g Copper oxychloride like Blue copper per litre of water can be beneficial.

» Treating seeds with 3 grammes of Captan or Thiram each kg of seed.

» Treating seeds with a formulation of *Trichoderma viride* weighing 4 grammes, along with 6 grammes of Metalaxyl, has proven to be quite successful.

» Captafol 0.2 %, Copper Oxychloride 0.25 %, and Carbendazim 1 % sprayed on.

» Applying Neem cake and *Trichoderma viride* to the soil.

» Prior to the commencement of the monsoon, spray a 0.2% solution of Metalaxyl or Mancozeb.

## Anthracnose or dieback, *Colletotrichum dematium*

## Symptoms

» Little grey patches on the leaves are a telltale sign.

» Depending on the severity, it might cause necrosis of twigs, lower branches, or even the entire top of the plant.

» All throughout the necrotic surface of the damaged twig, tiny black spots (acervuli) appear.

» The fruits eventually develop sunken areas ranging in colour from dark brown to black, often containing masses of pink spores.

» When the rot progresses inside the fruit, the seeds inside will rot and fall out.

» Chilli fruits with any sign of anthracnose diminish in value.

## Survival and spread

» Several Colletotrichum species store their acervuli and micro-sclerotia in and on seeds, where they can remain viable for years.

» Other hosts, such as other solanaceous or legume crops, plant debris, and rotting fruits in the field, are suitable for fungal overwintering.

» Conidia from acervuli and micro-sclerotia are spread from infected to healthy fruit and foliage when warm and moist conditions prevail.

» Infectious disease can be transmitted from plant to plant in a field by means of inoculum carried by diseased fruit.

## Management

» Plant chilies from pathogen-free seed.

» Every two to three years, you should switch up your chilli crop for something that isn't a food source for Colletotrichum.

» It is recommended to apply 2 g/kg of Carbendazim or 2 g/kg of Thiram 75 WP to the seed.

» Before sowing, seedlings should be treated with 0.1% Carbendazim.

» At every other spraying in the main field, spray 0.25 % Propineb or Mancozeb, 0.1 % Carbendazim or Thiophanate methyl, or 0.3 % Copper oxychloride.

» Azoxystrobin (Quadris), trifloxystrobin (Flint), and pyraclostrobin (Cabrio) are all Strobilurin fungicides that can be sprayed on chillies to prevent anthracnose.

» Capture infected vegetation and destroy it.

» Benomyl/Thiophanate methyl 1 g/liter sprays are effective in preventing the occurrence of post-harvest diseases.

**Verticillium wilt,** *Verticillium dahliae, V. albo-atrum*

**Symptoms**

» Stunting, defoliation, and wilting are outward signs of disease, as is darkening of the vascular system.

» Disease in the field typically manifests in the form of a single sick plant or a cluster of sick plants (foci).

» When the disease spreads from these initial points of infection, it has the potential to infest the entire plantation.

» The physiological state of infected plants may alter, maybe for the worse.

» When the leaves get brown, they fall off.

» Branches and stems wither and perish.

» When the roots and lower stem of a wilted plant are cut longitudinally, the vascular tissue inside turns a brownish tint.

**Favorable conditions**

» Cool air and soil temperatures are ideal for the growth of *Verticillium wilt* spores.

**Survival and spread**

» Without a host, *V. dahliae*-produced microsclerotia have the potential to persist in the wild for up to 14 years.

**Management**

» By heating the top 6 inches of soil to temperatures over the fungus's survival threshold, solarization effectively eradicates the problem.

» Little sick plants are simple to remove and destroy, and they can be replaced with new, healthy plants.

» Make sure to fertilize regularly with a fertilizer that has a low nitrogen content and a high phosphorus content.

» Eliminate any diseased or rotting branches.

» Reduce inoculum and subsequent plant infection by rotating with broccoli, corn, wheat, barley, sorghum, or safflower for at least 2 years (the longer the rotation, the better).

» *Verticillium* wilt can be mitigated by reducing pathogen populations in the soil with the application of Metham-sodium fumigants and the use of organic amendments.

## Bacterial wilt, *Ralstonia solanacearum*

### Symptoms

» The entire plant wilts without the leaves becoming yellow.

» If you cut a cross section of a diseased plant from the roots or lower stems and suspend it in water, you'll see milky streams of bacteria oozing out of the plant's vascular system.

### Management

» Use only disease-free seeds.

» Avoid planting in areas where bacterial wilt has occurred.

» If you want to keep your garden healthy, you should limit your use of nitrogen fertilizers.

» Take sick plants out of the field to stop the spread of disease.

» Get rid of the root-knot nematodes since they could potentially spread bacterial wilt.

» The bacterial wilt pathogen in soil can be inhibited by using soil amendments (organic manures).

» Planting solanaceous crops year after year is not advised.

» After a crop is harvested, all of the plant remnants should be incorporated back into the soil.

## Bacterial leaf spot, *Xanthomonas campestris* pv. *vesicatoria*

### Symptoms

» Chilli leaves, petioles, stems, and fruits develop tiny dark patches that are slightly elevated.

» These blemishes eventually become oily or dark.

» Tissue at the leaf margins dries up and cracks.

» Older leaves will be affected first, then the newer growth.

» At some point, leaves will fall off.

» There is a risk of brown cankers on stems.

» In the beginning, the fruit's swarms are soft and watery, but they soon harden and turn brown.

## Favorable conditions

» Hot and damp weather is ideal for the spread of disease.

## Management

» Plants should be disease-free.

» After harvesting, pick up and dispose of all crop trash.

» Submerge the diseased material by ploughing it under.

» Seeds are soaked in hot water at 50 °C for 25 minutes.

» Spraying with a solution of 3 parts water to 50 parts Bordeaux wine, 0.2 % Zineb, 20 parts per million of streptocycline, 100 parts per million of agrimycin and copper oxychloride, or 200 parts per million of paushamycin and copper oxychloride, is helpful.

» Pepper cv. Supposedly, Jwala can withstand the rigours of the field.

## Leaf curl disease, *Chilli Leaf Curl Virus*

## Symptoms

» Symptoms include vein thickening and swelling, interveinal puckering and blistering, and abaxial and adaxial leaf curling.

» The development of smaller clusters of leaves from axillary buds occurs at a later stage.

» Plant as a whole looks bushy due to its limited development.

» The flowers and fruits are scant and tiny, with twisted stems.

## Spread

» Acquired from aphids.

» Carbofuran soil treatment at 1.5 kg a.i. /ha in the nursery and again when you transplant, then two or three sprays with 0.05% Dimethoate or Monocrotophos at 10-day intervals.

## Management

» Removal of diseased plants from the field.

» The field and its surroundings must be cleared of weeds.

## References

Saini, A., Ahir, K. C., Rana, B. S., & Kumar, R. (2017). Population dynamics of sucking pests infesting chilli (Capsicum annum L.). *Journal of Entomology and Zoology Studies*, *5*(2), 250-252.

Pathipati, V. L., Vijayalakshmi, T., & Naidu, L. (2014). Seasonal incidence of major insect pests of chilli in relation to weather parameters in Andhra Pradesh. *Pest Management in Horticultural Ecosystems*, *20*(1), 36-40.

Kumar, V., Swaminathan, R., & Singh, H. (2015). Bio-efficacy of newer insecticides against sucking insect pests of chilli. *Annals of Plant Protection Sciences*, *23*(1), 69-73.

Hussain, F., & Abid, M. (2011). Pest and diseases of chilli crop in Pakistan: A review. *Int. J. Biol. Biotech*, *8*(2), 325-332.

Meena, R. S., Ameta, O. P., & Meena, B. L. (2013). Population dynamics of sucking pests and their correlation with weather parameters in chilli, Capsicum annum L. crop. *The Bioscan*, *8*(1), 177-180.

Thind, T. S., & Jhooty, J. S. (1985). Relative prevalence of fungal diseases of chilli fruits in Punjab. *Indian Journal of Mycology and Plant Pathology*, *15*(3), 305-307.

Sridhar, K., Rajesh, V., & Omprakash, S. (2014). A critical review on agronomic management of pests and diseases in chilli. *Int. J. Plant Anim. Environ. Sci*, *4*, 284-289.

Pandey, S. K., Mathur, A. C., & Manisha, S. (2010). Management of leaf curl disease of chilli (Capsicum annuum L.). *International Journal of Virology*, *6*(4), 246-250.

Berke, T., & Shieh, S. C. (2000). Chilli peppers in Asia. *Capsicum & Eggplant Newsletter*, (19), 38-41.

# 10

# Cowpea

## INSECT PESTS

### Spotted pod borer, *Maruca vitrata* (Syn. *M. testulalis*)

#### Crop losses

» Reported to cause a 70%-80% yield loss, according to reports.

#### Damage symptoms

» This bug lives in arid regions and can ruin crops by eating the seed.

» The larvae are hairy and olive green; they have rows of dark dots.

» Caterpillars do extensive harm to crops by feasting on buds and blossoms before boring into pods and devouring their seeds.

» Pods and blossoms infected with the pest web together.

#### Pest identification

» A single egg has the appearance of a tiny drop of water; it is oval in shape, transparent, and a pale yellowish white in colour.

» The caterpillar stage, or larvae, can reach lengths of 17-20 mm. Their bodies range from white to a pale green colour, and their dark brown heads are covered in irregular brown-black dots.

» The wings of an adult moth are 20-25 mm in size and have a blackish-brown base colour with white patterns on the forewings.

#### Management

» It's best to get your plants in the ground ahead of the peak pest activity.

» Choose early-maturing plant species.

» Crop cowpea with maize in rotation.

» Combining cowpeas with sorghum or corn as an intercrop.

» Pick the eggs and larvae out of the plants by hand and squash them.

» The Crotalaria species can be used as a trap crop.

» Neem cake at a rate of 250 kg per hectare should be applied throughout flowering, and the initial spray with pulverised Neem seed powder extract (NSPE) at flower bud formation is required.

» A 1% concentration of neem oil in soap. After a week, you should spray again.

» You can use Cypermethrin 25 EC at 0.5 ml/L, Chlorpyriphos 20 EC at 2.5 ml/L, or Indoxacarb 14.5 SC at 0.5 ml/L as sprays.

» Both the *Phanerotoma leucobasis* and the *Braunsia kriegeri* have natural enemies.

» The larvae can be killed by using *Bacillus thuringiensis* (Bt).

» Spotted pod borer is another insect host for MaviMNPV.

» Transgenic *Bt* cowpea has the potential to increase yields by up to 25% compared to conventional cowpea in Nigeria.

## Aphid, *Aphis craccivora*

### Damage symptoms

» Adults and juveniles consume plant sap from developing tips, immature foliage, and pods of mature plants.

» Plants with severe infestations became chlorotic and their leaves curl.

» Seedlings and plants infected with the pests die off quickly and their growth is inhibited.

» During its feeding process, this aphid secretes copious amounts of honeydew, which is then colonized by sooty mould. Black sooty mould inhibits plant photosynthesis.

» The spread of cowpea aphid-borne mosaic virus has far-reaching consequences, even in low numbers.

**Pest identification**

- » Black and lustrous, adults can grow to be 2 mm long, and some species even have wings.
- » A greyish and lifeless waxy layer covers a nymph's body.

**Management**

- » Set up sticky yellow traps. Yellow is a colour preferred by aphids.
- » Put some Neem seed kernel extract (NSKE, 5%) on the plants to keep the pests at bay.
- » Chlorpyriphos 2 ml/L in water or Dimethoate 30 EC 650 ml/ha in 600 L of water can be sprayed on young crops.
- » The following should be sprayed: Ambush, Baythroid, Furadan, Lorsban, and Mustang Max.
- » If pests are attacking the terminal buds of your young crop, spray it with Dimethoate at a rate of 1 litre per hectare 30 days after sowing.
- » *Menochilus sexmaculata* will be released at a rate of 1,250 per hectare.
- » Aphid-eating insects including ladybirds, lacewings, damsel bugs, and syrphid flies should be released.

**Blue butterfly,** *Lampides boeticus*

**Damage symptoms**

- » The larvae eat both the growing flower and the host plant's seed capsules.
- » The decaying of the pod may be hastened by the accumulation of frass at one end.
- » A noticeable external sign of this is frequently a dull, black color.
- » Bore holes in the beginnings of flowers and pods.
- » Observation of a caterpillar that looks like a slug.
- » Black ant activity accompanied with the emission of honey dew.

**Pest identification**

- » The larva is spherical and flat, with a pale green colour and a rough exterior.
- » The adult moth has a long, greyish blue tail and black markings on its hind wings; the underside of its wings is covered in stripes and brown spots.

## Management

- » By planting in early October, there was less damage to the seeds and pods.
- » Put up 50 birdhouses each hectare.
- » Placing light traps (1 trap/5 acres) to exterminate the pest population.
- » *Trichogramma chlionis* would be released at 1.5 lakh/ha/week for four weeks.
- » Save the green lacewing and other beneficial insects including predatory stink bugs, spiders, and ants.
- » The larval population was reduced most by *Purpureocillium lilacinum, Vetricillium lecanii,* and *Bacillus subtilis.*
- » Neem oil/Pongamia oil 1%, Neem seed kernel extract 4%, and 600 gms of *Bacillus thuringiensis* in a spray.
- » Twice with NSKE 5% and then once with Triazophos 0.05% spray.
- » Insecticides such as deltamethrin, carbaryl, and cyhalothrin can be sprayed.

## Blister beetle, *Mylabris phalerata*

## Damage symptoms

- » Worsen the situation by eating cowpea flowers.
- » Adult flower beetles, also called blister beetles, eat on flower buds and blossoms (petals and/or pollen), causing fewer pods to develop.
- » If insects invade in large quantities, the harvest could be ruined.
- » Damage to cowpea fields is common when they are adjacent to or intercropped with maize.

## Pest identification

- » The eggs are a pale yellow and have a spherical form.
- » Infant grubs are a dazzling white.
- » Adults' elytra are black with a white patch in the middle and two orange wavy lines running transversely across the wings.

## Management

- » It appears that the sole cure is to manually gather or catch with an insect net and then kill the adults in kerosene-treated water.

# DISEASES

## Damping-off, *Pythium aphanidermatum, Rhizoctonia solani*

### Symptoms

» Prematurely rotting seeds might kill off tender seedlings before they even emerge from the ground.

» Post-emergence damping off causes the collar region of the hypocotyl to develop reddish brown lesions, and then the young plants die quickly.

» If you have an infection caused by *R. solani*, the damage will only show up around your neck and shoulders.

» *P. aphanidermatum* has a lesion that spreads upward and a watery lesion that kills seedlings.

### Management

» Seeds must be confirmed as free of pathogens.

» Remove the surplus water from the playing field.

» *Trichoderma viride* 10g/kg, *Pseudomonas fluorescens* 10g/kg, Carbendazim 2g/kg, or Thiram 2g/kg seed treatment.

» Use 1 gramme per litre of Carbendazim or 2.5 kg of *Phytophthora fluorescens/ Trichophthora viride* per hectare of land using 50 kg of fresh wheat straw.

## Charcoal rot/ root rot, *Macrophomina phaseolina*

### Symptoms

» Stem discoloration near the ground.

» Upward canker spread is possible on a stem.

» The plant's leaves may wilt and fall off.

» Little, black sclerotia (the fruiting bodies of the fungus) form in the afflicted tissues, allowing for easy diagnosis.

» Some of the seedlings acquire rot in the hypocotyl area and continue to droop.

» Both the seedlings and the roots start to dry out.

» At maturity, cowpea plants develop depressed lesions on the lower stem and roots that have a greyish black colour and contain small black sclerotia.

» Plants can suffer from stem girdle, a condition that can cause the stem to split down its length.

## Spread

» In the majority of cases, the microsclerotia in the soil are responsible for the transmission of disease.

## Management

» Use of non-host crops in crop rotation for a minimum of two years.

» Farm yard manure or Neem cake can be added to the soil to help lower inoculum levels.

» Application of 2 grammes of Carbendazim or Thiram per kg of seeds.

» Seeds should be treated with *Trichoderma* or another biocontrol agent at a rate of 10g/kg of seeds, and biocontrol agents should be applied to the soil alongside organic manure.

## Anthracnose, *Colletotrichum destructivum*

## Symptoms

» The stems become affected, and brown, deep lesions with dark red edges appear.

» Several lesions quickly join together to encircle the main body of the plant, as well as any lateral stems, branches, peduncles, or petioles.

» Leaf surface lesions include tan-red patches with a yellow halo and elongated lesions along leaf veins.

» Pink spore masses may form on lesions during moist periods.

» Pods can get lesions too, just like the stem does.

» Affected pods fold and lack normally sized seeds.

## Spread

» As a result of splashing rain and rain blown by the wind, fungus spores might spread.

## Management

» Never risk planting contaminated seedlings; stick to the tried and true methods.

» To lessen the chance of inoculation, it is important to maintain proper field sanitation by clearing away crop debris after harvest.

» Use of non-host crops in crop rotation for a minimum of two years.

» Combining cereal crops with other plant types.

» Application of 2 gms of Carbendazim or Thiram per kg of seeds.

» Every two weeks, spray either 0.2% Mancozeb or 0.2% Carbendazim.

## Cercospora leaf spot, *Cercospora canescens*

## Crop losses

» Crop loss due to *Cercospora* was over 40 %, according to field observations.

## Symptoms

» Typically, necrotic lesions of varying sizes and shapes, ranging in colour from bright red to a rusty brown, appear on both sides of the affected leaves.

» Leaves are turning yellow.

» When spores are made, the centre turns a silvery grey.

» Lesions, which look like black matting on the underside of the leaf, are covered in spores.

» There may be defoliation if the disease is particularly severe.

## Survival and spread

» The pathogen is able to thrive in infected seed, infected crop debris, and alternate hosts.

» The spores are dispersed by the wind and rain.

## Management

» After harvesting, clear the field of all agricultural remains.

» Grow plants from seed that hasn't been contaminated with disease.

» Before seeding your field, clear off any stray cowpea or legume plants.

» Cowpea (or other legume) intercropping with corn (or sorghum) or wheat.

» Application of 2 gms of Carbendazim or Thiram per kg of seeds.

» The disease was most effectively managed by applying Benomyl spray once a week, beginning three weeks after planting.

» Every two weeks, spray either Mancozeb (0.2%) or Carbendazim (0.2%).

**Septoria leaf spot,** *Septoria vignae*

**Symptoms**

- » Little, pinhead-sized, reddish-brown, circular, water-soaked dispersed dots occur on both sides of the leaves.

- » Spots quickly develop into larger, asymmetric lesions.

- » After some time, spots turn a pale brown in the centre and a darker shade around the edges.

**Management**

- » Every two weeks, spray either Mancozeb (0.2%) or Carbendazim (0.2%).

**Brown rust,** *Uromyces appendiculatus*

**Symptoms**

- » First, the lower leaves begin to turn yellow and wilt.

- » The leaves have tiny pustules of a reddish brown colour.

- » As time goes on, the upper leaves succumb to the disease.

- » There are little reddish brown pustules on the stems, and occasionally you'll see white hyphae sticking out.

- » When pods do form, they tend to be small, have brown markings, and show sporulation.

- » Deformed seeds have a harder time germinating.

- » The leaves on the plant are drying up and falling off.

- » Rust causes a browning effect on infected plants.

- » In nature, we see stunting and weak, pale green growth.

**Favorable conditions**

- » When the weather is hot and wet, fungi can grow quickly.

- » Overhead watering, seedling crowding, and a lack of ventilation all contribute to the spread of the disease.

**Survival and spread**

- » Infected plant detritus or even structural elements might serve as a safe haven for fungi to overwinter.

» Transplants or seeds could also be contaminated.

## Management

» The practises of cleaning up the area, thinning the seedlings, weeding, and rotating crops every four to five years.

» Avoid using seed from diseased farms and instead seek out sources of disease-free seed.

» At the first sign of disease, pull out any affected plants and promptly spray the remaining crop.

» Fungicides like Mancozeb can be applied to seeds before they are planted.

» Every two weeks, spray either Mancozeb (0.2%), Oxycarboxin (0.1%), or Chlorothalonil.

» Sulfur or potassium carbonate sprays might be used to help contain the disease.

## Bacterial blight, *Xanthomonas campestris*

## Symptoms

» Infected areas on leaves swell and becoming necrotic from constant water exposure.

» In certain cases, spots will have a yellow discoloured ring around them.

» The coalescence of lesions gives the plant a burned appearance, and dead leaves do not fall off the plant.

» Pods may have a round, depressed, reddish-brown lesion.

» In moist environments, pod lesions might leak.

## Favorable conditions

» Warm weather facilitates the spread of diseases.

## Survival and spread

» Infectious diseases can be spread through planting seed that has been tampered with.

» Crop debris serves as a winter home for bacteria.

» Conditions of high humidity and precipitation promote the most rapid spread.

## Management

> » Only use certified seed in your plantings.

> » Cowpeas (edible or soybeans) should not be planted in the same field more frequently than once every three years.

> » Produce hardy seedlings. Modern cowpea varieties that are resistant include OAC Rex, Lighthouse, Mist, Apex, and Rexeter.

> » To prevent the spread of bacteria during planting, seeds should be treated with the appropriate antibiotic. Seeds treated with streptomycin will have much little or no surface contamination after harvest.

> » As soon as symptoms are detected, spray the plants with a fungicide that contains copper.

### *Cowpea mosaic virus*

## Symptoms

> » Result in a random mosaic of bright and dark green on the leaves.

> » Result in leaves that are thick, abnormal, and deformed.

> » Younger leaves are the finest for seeing the mosaic patterns.

> » It's possible that the plants won't grow properly and won't have any pods.

> » Pods won't form if the plant gets sick while it's still young.

## Spread

> » The use of disease-resistant plant cultivars should be encouraged.

> » Plant only disease-free seeds.

> » Non-leguminous crops should be rotated in for four to five growing seasons.

> » It's imperative to promptly get rid of any contaminated plants.

> » It's also important to get rid of weeds and any other alternate hosts.

> » Prevention of secondary infection through the control of vectors through the use of pesticide spray. Spraying insecticide, such as 0.1% Chlorpyriphos, can help eliminate the pest.

# References

Adegbite, A. A., & Amusa, N. A. (2008). The major economic field diseases of cowpea in the humid agro-ecologies of South-western Nigeria. *African journal of Biotechnology*, 7(25).

Booker, R. H. (1965). Pests of cowpea and their control in Northern Nigeria. *Bulletin of Entomological Research*, *55*(4), 663-672.

Emechebe, A. M., & Lagoke, S. T. O. (2002). Recent advances in research on cowpea diseases. *Challenges and opportunities for enhancing sustainable cowpea production*, *5*, 94.

Jackai, L. E., & Daoust, R. A. (1986). Insect pests of cowpeas. *Annual review of entomology*, *31*(1), 95-119.

Singh, S. R., & Allen, D. J. (1979). Cowpea pests and diseases. *Manual*.

Togola, A., Boukar, O., Belko, N., Chamarthi, S. K., Fatokun, C., Tamo, M., & Oigiangbe, N. (2017). Host plant resistance to insect pests of cowpea (Vigna unguiculata L. Walp.): achievements and future prospects. *Euphytica*, *213*, 1-16.

Singh, S. R. (1978). Resistance to pests of cowpea in Nigeria. *Pests of grain legumes: Ecology and control*, 267-279.

Singh, B. B. (Ed.). (1997). *Advances in cowpea research*. IITA.

Singh, B. B., Chambliss, O. L., & Sharma, B. (1997). Recent advances in cowpea breeding. *Advances in cowpea research*, 30.

Williams, R. J. (1975). Diseases of cowpea (Vigna unguiculata (L.) Walp.) in Nigeria. *PANS Pest Articles & News Summaries*, *21*(3), 253-267.

# 11

# Curry Leaf

## INSECT PESTS

**Psylla,** *Diaphorina citri*

### Damage symptoms

> » Curry leaves are a favourite food of citrus psyllids.

> » This can bring bacteria to the tree, in addition to causing damage to the leaves and stems.

> » Leaf curling and twisting, as well as shoot death, are symptoms.

> » Honeydew, a sticky secretion from nymphs, is a prime food source for sooty mould.

### Pest identification

> » Flat-bodied, yellow, orange, or brown nymphs. Their tiny size (approximately 1/100 to 1/14 of an inch) makes them difficult to spot.

> » Adults are a tiny, brown-and-black speckled bug measuring only 1/16 to 1/8 of an inch in length, with red eyes and short antennae.

### Management

> » The infected areas of the plant should be closely monitored and pruned.

> » Mineral oils for gardening, diluted to 0.5–1.0% and sprayed

> » Dimethoate sprayed at 1 ml/ lit is effective against psyllid bugs and scales.

> » Predatory wasps, lacewings, lady beetles (*Chilocorus nigritus*), lacewing larvae, syrphid fly larvae, predatory ants, predatory mites, etc. should be protected.

» The nymphs of parasitic wasps, such as *Tamarixia radiata*, *Diaphorencyrtus aligarhensis*, etc., are examples of parasitoids.

## Lemon butterfly, *Papilio demoleus*

### Damage symptoms

» Caterpillars devour young leaves up to their midribs, killing entire trees or seedlings and leaving just the trunks and branches.

### Pest identification

» Eggs are placed singly on young leaves and shoots and are a pale yellowish white in colour.

» Caterpillar larvae are initially a dark brown colour, but quickly acquire white, splotchy markings that resemble bird's drop.

» Adults are a large, stunning butterfly, measuring between 50 and 60 mm across its wings. There is no tail-like projection behind it, and its hind wings contain a brick-red oval patch towards the anal border.

### Management

» Killing the larvae by hand and spraying them with Malathion (1 ml/liter) are both viable options.

» The use of 4-5 yellow sticky traps per acre for monitoring purposes is recommended.

» Mineral oils for gardening, sprayed at a concentration of 0.5-1.0%

» Cow dung ash used as a dusting.

» Azadirachtin sprayed in concentrations of 5% NSKE or 0.03%

» When applied at a concentration of 1 part per 25 parts water, fish oil rosin soap is very helpful in combating these bloodsucking pests.

» Retain predators like wasps, ladybirds, lacewings, and syrphid fly larvae.

» *Chrysoperla carnea*, coccinellids, king crow, common mynah, wasp, dragonfly, spider, robber fly, reduviid insect, praying mantis, fire ants, huge eyed bugs (*Geocoris* sp.), pentatomid bug (*Eocanthecona furcellata*), earwigs, ground beetles, rove beetles, etc.

» Parasitoids can take several forms, including the eggs of *Trichogramma* spp. and *Telenomus* sp., as well as the larvae and pupae of *Distatrix papilionis*, *Brachymeria* spp., *Pteromalus* spp., and others.

## Scale insect, *Unaspis citri*

### Damage symptoms

» Plants suffer harm from scales because they drain the plant's sap, causing the leaves to turn yellow and wilt.

» A single scale won't kill a curry tree, but if this insect is allowed to multiply, it can quickly become a problem.

» Constant vigilance can also detect scaling issues on the curry tree before they become serious infestations.

### Management

» Dimethoate sprayed at 1 ml/ lit is effective against psyllid bugs and scales.

» The nematode parasite *Aphytis melinus* belongs to the phylum Parasitoda.

» Mealy bug killers (a conccinellid, *Cryptolaemus montrouzieri*), predatory wasps, syrphid/hover flies, ladybirds, lacewings, ants, and so on.

## Mealy bugs, *Planococcus citri*

### Damage symptoms

» Yellow nymphs, which feed on plant sap, emerge from mealy bug eggs, which are laid in clusters of several hundred on the surface of a leaf.

» Large populations of citrus mealy bugs can cause not just leaf withering and fall, but also fruit drop in infested trees.

» Mealy bugs tend to congregate in great numbers, and their feeding causes leaves to fall off and branches to die off before they should.

» The honeydew they produce is food for black sooty mould, much as it is for psyllids.

### Pest identification

» The waxy insects known as citrus mealy bugs have a rosy white look.

» Mealy bugs are little, wingless insects that measure between 1/20 and 1/5 of an inch in length.

### Management

» Mealy bugs are a common host for parasitic wasps.

» Predators include midges, green lacewings, ladybird beetles or mealy bug destroyers (Coccinellid, *Cryptolaemus montrouzieri*), syrphid/hover flies, wasps, and more.

## Aphids, *Toxoptera aurantii*

### Damage symptoms

» When aphids feed on a plant, the leaves will curl and mottle as they extract nutrients.

» Possible fungal/mold introduction as well.

» During feeding, aphids often congregate in large groups and move slowly in response to disturbances.

### Pest identification

» Aphids are tiny insects that can be any one of several colours and typically have a pear shape.

### Management

» Protect parasitoids like the aphid-eating *Aphidius colemani*, earwigs like *Diaeretiella*, and ants like *Aphelinus*.

» Protect predators such anthocorid bugs (*Orius* spp. ), mirid bugs, syrphid/hover flies, green lacewings (*Mallada basalis* and *Chrysoperla carnea*), predatory coccinellids (*Stethorus punctillum*), staphylinid beetles (*Oligota* spp.), predatory cecidomyiid fly (*Aphidoletis aphidimyza*) and predatory gall midge, etc.

## Two spotted spider mite, *Tetranychus urticae*

### Damage symptoms

» Stippling, scarring, and a bronzing of the leaves and calyx are symptoms of damage by two-spotted spider mites and carmine spider mites.

» Problems occurring between two and five months after a late-summer or fall transplant.

» With infection rates higher than one mite per leaflet, yield loss becomes obvious.

### Pest identification

» Six-legged nymphs are produced in the first stage, and nymphs with eight legs are produced in the second.

**Favorable conditions**

» The generation time might be as little as four days in warm climates.

**Survival**

» These mites, known as two-spotted mites, spend the winter underground.

**Management**

» Anthocorid bugs (*Orius* spp.), mirid bugs, syrphid/hover flies, green lacewings (*Mallada basalis* and *Chrysoperla carnea*), predatory mites (*Amblyseius alstoniae*, *A. womersleyi*, *A. fallacies*, and *Phytoseiulus persimilis*), predatory coccinellids (*Stethorus punctillum*), staphylinid beetle (*Oligota* spp.), predatory cecidomyiid fly (*Anthrocnodax occidentalis*), predatory gall midge (*Feltiella minuta*), etc.

» *Beauveria bassiana* is a pathogenic microorganism (entomopathogen).

# DISEASES

*Phyllosticta* **Leaf spot,** *Phyllosticta* **sp.**

**Symptoms**

» The *Phyllosticta* leaf spot symptoms include a few small, circular blemishes on the leaves.

» Extreme infestations can weaken trees to the point where they lose leaves prematurely.

» The lesions appear on the leaves as small, circular, yellowish-brown spots.

» Toxic fungi grow small black fruiting bodies in the shape of a circle under optimal conditions.

» These spots have dead tissue at their centres, which can be easily removed.

**Spread**

» Soil and rain splashes are the primary vectors of infection.

» When conditions are damp, secondary infection can spread through the air and rain splash.

**Management**

» Always plant from disease-free seed or use seedlings.

» A preventative spray made from a Tobacco decoction could be used.

» Maximum fungal growth inhibition was achieved with 1360 ppm citrus oil, then 1720 ppm lemongrass oil, and 2260 ppm peppermint oil.

» 5% NSKE spray.

» Carbendazim at a concentration of 1 mg/ ml of water is effective for spraying on plants to prevent the spread of leaf spot disease. Do not use a spray that contains sulphur compounds.

### *Colletotrichum* leaf spot, *Colletotrichum gloeosporioides*

### Symptoms

» First manifesting as brownish to black spots on the stem, symptoms spread higher.

» This disease causes a blackening of the veins in damaged stem branches and leaves, as well as tissue necrosis.

» Lesions on the leaves ranged in shape from circular to irregular, depressed in the centre, and bordered by yellow edges.

» Acervuli and black setae were dispersed throughout the stomatic tissues of leaves and young branches.

### Favorable conditions

» Favored conditions include lots of moisture in the air.

### Survival and spread

» By developing appressoria, pathogens are able to more easily invade their hosts, and these projections are also thought to contribute to the pathogen's long-term viability.

» Wind and rain carry the conidia to other parts of the plant.

### Management

» Spraying a fungicidal solution (0.1%) of Bavistin on the plant's foliage has been shown to be beneficial in controlling the disease.

### *Cercospora* leaf spot, *Cercospora* sp.

### Symptoms

» Damages curry leaves and makes them unusable.

## Management

» Carbendazim, sprayed at a concentration of 1 mg/ ml of water, is an effective method for controlling leaf spot disease.

## References

Devaki, K., Muralikrishna, T., Sreedevi, K., & Rao, A. R. (2012). Incidence and biology of leaf roller, Psorosticha zizyphi (Stainton)(Lepidoptera: Oecophoridae) on curry leaf, Murraya koenigii (L.) Sprengel. *Pest Management in Horticultural Ecosystems*, *18*(2), 154-157.

Gogikar, P., Vemuri, S., Reddy, C. N., & Senivarapu, S. (2017). Dissipation pattern of triazophos and chlorpyriphos in curry leaf. *Open Access Library Journal*, *4*(1), 1-8.

Ramakrishnan, N., Sridharan, S., & Chandrasekaran, S. (2015). Insecticide usage patterns on curry leaf. *International Journal of Vegetable Science*, *21*(4), 318-322.

Sahu, C. M., Nirala, Y. P. S., Painkra, K. L., & Ganguli, J. L. (2015). " Seasonal incidence of Citrus Butterfly, Papilio demoleus Linnaeus (Lepidoptera: Papilionidae) on Curry leaf, Murraya koenigii at Raipur (CG)". *International Journal of Tropical Agriculture*, *33*(2 (Part I)), 525-528.

Tara, J. S., & Sharma, M. (2010). Survey of Insect pest diversity on economically important plant, Murraya Koenigii (l.) Sprengel in Jammu, J&K. *Journal of Entomological Research*, *34*(3), 265-270.

# 12

# Drumstick

## INSECT PESTS

### Bud worm, *Noorda moringae*

### Damage symptoms

- » In their early stages, caterpillars form large groups and feed on the leaf lamina, reducing the leaf's chlorophyll content and leaving the leaf looking papery white.
- » Later, they develop into ravenous feeders that eat holes in the leaves in an erratic pattern.
- » Veins and petioles are all that's left of the leaves after they've been subsequently skeletonized.
- » Massive defoliation takes place.
- » Boring fruits have been spotted with weird holes in them.
- » Insect larvae, which eat on the developing flower, can reduce a summer bloom by as much as 78%.
- » A single caterpillar is observed in a flower bud.

### Pest identification

- » Eggs are placed individually or in small clusters on flower buds, and are a creamy oval shape.
- » The larvae have a dirty brown colour with a black head and prothoracic shield.
- » A cocoon made of soil is used during the pupation process.
- » Both the front and back wings of an adult butterfly are a pale, brownish yellow.

**Favorable conditions**

» When the weather warms up in the summer, the insects in southern India become more active.

**Management**

» It is necessary to plough around trees in order to reach the pupae and kill them.

» Remove the caterpillars and any damaged buds at the same time.

» Attract and kill adults using light traps at a rate of 1-2/ha.

» Use Carbaryl 50WP at 1 kg/ha or Malathion at 1 liter/hectare in a spray volume of 500-750 liters/hectare.

**Leaf caterpillar,** *Noorda blitiealis*

**Damage symptoms**

» Caterpillars eat the lamina of leaves, transforming them into paper-thin parchment.

» Initiate a defoliation.

» Infestations are worst in the months of March–April and December–January.

**Pest identification**

» Oval, white, creamy eggs are placed in bunches on the undersides of leaves.

» Larvae do not have a prothoracic shield.

» The adult stage resembles the budworm but is much larger.

**Favorable conditions**

» The months of March and April, as well as December and January, are the most infested.

**Management**

» It is necessary to plough around trees in order to reach the pupae and kill them.

» Get rid of the caterpillars and any damaged buds at the same time.

» The optimal density for light traps is 1-2/ha.

» You can use Malathion 50 EC at 2 ml/L or Carbaryl 50 WP at 1 g/L to spray the area.

**Hairy caterpillar,** *Eupterote mollifera*

**Damage symptoms**

- » During the hottest parts of the day, the caterpillars congregate in groups on the plants' stems.
- » Caterpillars scrape bark and eat foliage in a communal feeding behaviour.
- » Nighttime predators, they quickly strip a tree of its leaves and cluster there.
- » In extreme cases of infection, the tree may lose all of its leaves.

**Pest identification**

- » Wings of the adult moth are a pale golden brown and are marked with thin lines.
- » Larvae are a dirty brown tint with tufts of white hairs emerging from little warts. Dangerous and unpleasant hairs are everywhere. Red coral coloration is seen on the head capsule and thoracic legs.

**Management**

- » The larvae are killed in groups as they crawl down the tree trunks, using a torch or flame thrower to scorch them.
- » The egg masses and caterpillars must be collected and disposed of.
- » Light traps should be placed at a density of 1-2 per hectare to attract and kill adults soon after rains.

**Pod fly,** *Gitona distigma*

**Damage symptoms**

- » Maggots gain access to soft fruits by boring tiny entrances in their tips.
- » The outcome is that the fruits dry out from the bottom up as the sticky juice oozes out.
- » Maggot populations on fruits rarely exceed twenty to twenty-eight.
- » The fruit's inner workings decay away.

**Pest identification**

- » The adult form of this fly is bright yellow and has bright red eyes.
- » There are maggots, and they have a creamy tint.

» Eggs are cigar-shaped and are deposited in clusters on the grooves of young pods.

## Management

» Remove all spoiled and damaged fruit off the ground and throw it away.

» Citronella oil, eucalyptus oil, vinegar (acetic acid), dextrose, or lactic acid are all effective attractants that can be used to lure in the adults.

» Destroy puparia by raking up the ground beneath trees or ploughing the affected area.

» When fruit set has reached 50%, spray NSKE 5% again 35 days later.

» At fruit set, spray 500-750 L of water with 500 ml of Dichlorvos 76 SC or 750 ml of Malathion 50 EC. Repeat the spraying 35 days later.

### Long haired beetle, *Batocera rubus*

### Damage symptoms

» The grubs build wavy tunnels under the bark, devour the underlying tissue until they reach the sapwood, and eventually kill the branch or stem.

» The bark of young twigs and petioles is a staple food for adults.

### Pest identification

» Grubs are chubby, measuring around 10 cms in length, being a yellowish tint, and having well delineated body segments.

» Beetles of this species reach adulthood with a yellowish brown body and white dots on the elytra.

### Management

» Remove any waste, webs, etc., from the afflicted area of the tree.

» To remove pests, fill each hole with cotton soaked with Monocrotophos 36 WSC 5 ml or another effective fumigant such as carbon disulphide, carbon tetrachloride, chloroform, or even gasoline, and then close the holes with mud.

# DISEASES

## Damping off, *Pythium debarryanum*

### Symptoms

» Infected seedlings and nursery soil lead to low crop yields.

» A staggering 25–75% death rate among seedlings.

» The seeds don't germinate because the seedlings die before they emerge from the earth (Pre-emergence damping off).

» Post-emergence damping-off is characterised by the appearance of disease in seedlings after they have emerged from the soil but before their stems have lignified.

» The collar area develops a lesion after being submerged in water.

» Browning and decay characterise infected areas.

» Plants wither and die because their tissues become mushy.

### Favorable conditions

» Continual, relentless, and heavy rain.

» Watering that is both copious and constant.

» Soil drainage issues and narrow planting distances.

» Warm (about 25-30°C) soil with a high moisture content.

### Survival and spread

» Its first source is soil oospores.

» Zoospores in the irrigation water is a secondary concern.

### Management

» A well-drained, light soil is ideal for growing a nursery.

» Farm waste is being burned on top of the beds.

» Planting seeds in pots or 15-centimeter-high raised beds.

» Plant at a sparse 650 g/40 sq. m.

» Seeds, seedlings, planting material, and soil should all be treated with *Trichoderma viride/harzianum* and *Pseudomonas fluorescens*.

# References

Chandrakar, T., & Gupta, A. K. (2020). Seasonal incidence of insect pests on drumstick (Moringa oleifera Lamk.). *Journal of Entomology and Zoology Studies, 8*(4), 1384-1387.

Debnath, S., & Nath, P. S. (2002). Management of yellow vein mosaic disease of okra through insecticides, plant products and suitable varieties. *Annals of Plant Protection Sciences, 10*(2), 340-342.

Ekabote, S. D., Hosagoudar, K., Gowda, P., & Sreenivasa, M. Y. (2023). First report of Fusarium incarnatum associated with fruit rot disease of drumstick (Moringa oleifera L.) in India. *Plant Disease*, (ja).

Halder, J., & Rai, A. B. (2014). New record of leaf caterpillar, Noorda blitealis Walker (Lepidoptera: Pyralidae) as fruit and seed borer of drumstick, Moringa oleifera Lam. *Journal of plant Protection and Environment, 11*(2), 6-9.

Kshirsagar, C. R., & D'Souza, T. F. (1989). A new disease of drumstick. *Journal of Maharashtra Agricultural Universities, 14*(2), 241-242.

Kashyap, P., Kumar, S., Riar, C. S., Jindal, N., Baniwal, P., Guiné, R. P., ... & Kumar, H. (2022). Recent advances in Drumstick (Moringa oleifera) leaves bioactive compounds: Composition, health benefits, bioaccessibility, and dietary applications. *Antioxidants, 11*(2), 402.

Kotikal, Y. K., & Math, M. (2016). Insect and non-insect pests associated with drumstick, Moringa oleifera (Lamk.). *Journal of Global Biosciences, 5*(4), 3902-3916.

Mahesh, M., Kotikal, Y. K., & Madalageri, M. B. (2013). Studies on natural enemies of insect pests of drumstick. *Journal of Biological control, 27*(4), 336-339.

Nambiar, V. S., Guin, P., Parnami, S., & Daniel, M. (2010). Impact of antioxidants from drumstick leaves on the lipid profile of hyperlipidemics. *J Herb Med Toxicol, 4*(1), 165-172.

Suresh, K., Rani, B. U., & Baskaran, R. M. (2022). Pests and Their Management in Drumstick. *Trends in Horticultural Entomology*, 1163-1175.

# 13

# French Bean

## INSECT PESTS

### Bean fly, *Ophiomia phaseoli*

#### Damage symptoms

- » The larvae feed on the lamina, veins, midrib, petiole, and stem of the leaf.
- » The larvae cause damage to seedlings by feeding on the stem, which causes the plant to expand and split just above the soil.
- » Stem/plant withering occurs after feeding.
- » Plants that have been damaged turn yellow, wilt, and become stunted; without adventitious (secondary) roots, they may not survive.
- » Puncturing a leaf allows the female to deposit her eggs beneath the epidermis, where they develop into the white spots that are commonly misidentified as signs of sickness.
- » Within the first few days after a seed germinates, it's common to find these dried ovipositional chambers.

#### Pest identification

- » Eggs can be either opaque or translucent, and they have an oval shape.
- » The larvae in the second instar are a dingy, off-white colour. The third-stage larva has a pale-yellow color.
- » The pupa has a brownish yellow color and a barrel form. The puparium turns a deep brown just before the adult emerges.
- » The adults of this species of fly have a shiny metallic black body, light brown

eyes, and hyaline wings with a noticeable notch towards the margin.

## Management

> After three days, when the cotyledons have emerged from the soil, earth up the plants.

> When grown alongside beans, onions can help reduce the number of bean flies.

> The adults of the bean fly deposit eggs on moth bean, chickpea, lentil, and cluster bean, but the eggs do not hatch.

> Spraying 0.07% Endosulfan or 4% NSKE/4% crushed NSPE/1% Neem soap/1% Pongamia soap on unifoliate leaves as soon as puncture scars from adult oviposition inside the leaf lamina are visible is recommended. After 15–20 DAS, administer a second spray.

> Carbofuran 3G applied to the soil at a rate of 15 kg/ha just before planting.

> Apply 200 kg per hectare of Neem cake.

> Parasitoid flies, specifically *Opius phaseoli* (Hymenoptera: Braconidae), are common.

## Aphid, *Aphis craccivora*

## Damage symptoms

> Seedling leaves, stems, and developing pods are some of the most popular aphid targets (late-season attack).

> Colonize plant parts (such as stems, leaves, and growth tips), feed on plant sap (resulting in plant wilt and death), then spread by seed.

> While heavy populations can cause leaf distortion and rolling due to feeding, plants are often fine with this level of damage.

> Addiction to pod food is a serious issue. The lack of sap slows pod development, and irregular cell growth may result.

> In the worst situations, plants may be stunted and their development may be slowed.

> Due to aphid infestation, the growth of older plants may be stunted.

> *Bean Common Mosaic Virus* is transmitted via them (BCMV)

## Pest identification

» *A. craccivora* has a dark green coloration.

## Favorable conditions

» Aphid populations tend to rise in dry conditions, making infestations worse during that time of year.

## Management

» Aphids can be trapped with ease using yellow sticky traps.

» Remove diseased branches and spray with a 1% solution of Neem or Pongamia soap.

» In the field, a single spray of a water suspension of *Fusarium pallidoroseum* ($7 \times 10^6$ spores/ml) was sufficient to effectively suppress aphid populations.

» Quinalphos 0.05% sprayed plants still experienced reinfestation and required additional insecticide applications (3 sprays).

» Acephate 75 WP at 0.75 ml/L or Dimethoate 30 EC at 2 ml/L can be sprayed.

## Serpentine leaf miner, *Liriomyza trifolii*

## Damage symptoms

» The larvae tunnel through the cell walls of leaves to access the dripping sap in the stem.

» Due to the larvae's eating in the stem, the plant may die or suffer stunted development.

» Young cotyledons and leaves have been severely damaged.

» In the lamina of a leaf, maggots will chew a trail that looks like a serpent.

» Plant mortality is guaranteed at high and early incidence.

## Pest identification

» The full length of an adult is less than 2 mm, and its wingspan is just 1.25 to 1.9 mm. The eyes on its yellow head are red. Gray and black predominate on the dorsal surface and thorax, while the ventral surface and legs are yellow. There is no covering on the wings; they are see-through.

## Management

> » Neem cake at the rate of 250 kg per hectare should be applied right after planting. Don't linger, especially now during the kharif season.

> » Early observation of the plants for signs of adult activity, such as puncture marks and petiole mining, is critical.

> » As soon as you spot even a few adults flying around your crop, spray it with either Endosulfan 35 EC at 2 ml/L, Acephate 75 WP at 0.75 g/L, NSPE at 5%, Neem soap at 1%, or a Neem formulation containing 10,000 ppm of azadirachtin.

> » Petiole mining is treated with a second spraying when, on average, 5 out of 10 leaves show symptoms (about 12 - 20 days of sowing).

> » Use of 4% NSKE at 10 and 15 DAS At 21 DAS, the pod production of French bean was 5.05 t/ha, up from 1.05 t/ha in the control, and the frequency of serpentine leaf miners was 15.20, a significant decrease from the 62.93 per cent in the control.

## Whiteflies, *Bemisia tabaci, Trialeurodes vaporariorum*

## Damage symptoms

> » When adults and larvae pierce leaves to syphon sap, it can stunt growth, turn leaves yellow, and even kill the plant.

> » Cause growth issues in plants by secreting honeydew, which can promote the development of sooty mould on leaves and pods.

> » Infested bean pods with black soot mould are worthless.

## Pest identification

> » Eggs are a pear form, with a diameter of around 0.2 mm; they are initially white but gradually darken to a brown.

> » Its adults are only a few mm in length, and its two sets of wings are carried in a rooflike position over its body. Those little insects look like tiny moths.

## Favorable conditions

> » It thrives in humid environments and is at its most destructive during the dry season.

## Management

- » Protect the earth's natural foes. Whitefly population management relies heavily on parasitic wasps.

- » The crop should be sprayed with Neem oil whenever possible. Pesticides derived from the neem tree have been found to diminish egg laying by adult whiteflies and prevent further development of the pests in their juvenile stages.

- » Parasitic hymenopteran insects of the genus *Encarsia* are thought to be effective non-chemical alternatives to pesticides.

## Red spider mite, *Tetranychus cinnabarinus*

## Damage symptoms

- » Infested plants may be debilitated or die off entirely as a result of the larvae's feeding and the subsequent production of excessive webs.

- » Brown spots on the underside of bean leaves are a telltale sign that they've been fed on.

- » White or grey spots appear on the underside of leaves due to webbed colonies of spider mites.

- » When the pest population is large, leaves wither.

- » Reduce plant development and flowering, pod production and length, and seed production per pod.

- » Mites cause the most harm to young plants.

## Pest identification

- » The body of an adult female is roughly oval in shape and about 1/50 of an inch in length. When compared to females, guys are narrower and shorter. They are mostly colourless save for a pair of black spots on their sides.

- » Light pink eggs.

## Favorable conditions

- » Extreme damage could occur during the dry season.

## Management

- » Keep away from using broad-spectrum insecticides, especially Pyrethroids, as they might cause spider mite infestations.

» To get rid of mites and their webs, spray plants with a powerful jet of water or use overhead watering.

» Apply Wettable Sulfur 50 WP (2 g/L), Propargite 57% (2.2 ml/L), or Fenazaquin 10% (2.2 ml/L) in water.

» Inject 10 mature *Phytoseilus persimilis* predatory mites into each plant.

# DISEASES

**Root rot, *Fusarium solani*, *Rhizoctonia solani* and *Pythium* sp.**

## Symptoms

### *Fusarium solani*

» Typically, plants will be stunted or become yellow, but they will not die.

» Red lesions, initially brown or black, can be seen on the taproot and lower stem.

» The reddish-colored tip of the taproot and any lateral roots could rot, wither, and die.

» Above the lesion, rootlets could grow to help the plant survive.

### *Pythium* root rot

» Long, wet patches can be noticed on the hypocotyls and roots of some plants.

» Because to the moist soft rot, these spots become slightly sunken and covered with tannish-brown lesions that cluster, giving the entire root system and lower stem a collapsed, shrunken appearance.

» When a plant's primary and secondary roots rot, it becomes severely stunted or dies.

### *Rhizoctonia solani*

» Damping-off of seedlings and rotted seeds.

» Older plants are stunted, wilting, and dying.

» On the roots and lower stems, lengthy, deep, reddish-brown lesions appear at or below the soil surface.

» Sick plants may get dwarfed, have yellowing leaves, and eventually die.

» Never allow your plants to dry out completely.

» There are bean kinds that can handle heat better than others.

## Survival

» Several years can pass without the fungus dying in the soil.

## Management

» After harvesting beans, bury the leftover material by ploughing it under.

» Eliminate legumes from your crop rotation every 6-8 years.

» The best way to grow beans is in ridges or hills.

## Anthracnose, *Colletotrichum lindemuthianum*

## Crop losses

» It is possible for yield losses to reach 100% if contaminated seeds are planted and the environment is right for the disease to flourish.

## Symptoms

» On the undersides of the leaves, along the leaf veins, linear, dark brick-red to black lesions can be seen.

» The upper leaf surface is the first to become discoloured as the disease advances.

» Some of the most noticeable signs appear on the pods.

» Spots ranging in colour from reddish brown to black and conspicuous circular lesions also appear frequently on pods.

» Most mature lesions have a grayish-black centre and a round, reddish-brown to black border.

» During wet times, the lesion's inside may leak pink spore masses.

» Pods that are diseased to a severe degree may shrivel and their accompanying seeds will likely be contaminated as well.

» Sunken sores and dark to black spots characterise infected seeds.

## Favorable conditions

» In order for the fungus to get established, there must be a period of sustained moisture.

» Regular precipitation has been linked to an increase in the prevalence and severity of several diseases.

## Survival and spread

» It is possible for fungi to overwinter in plant remains in the soil and re-infect crops the next growing season.

» Spots produce spores, which are carried by rain splash or the combination of wind and rain.

» Fungus propagates from infected seeds and can travel great distances.

## Management

» Sow only certified disease-free seed.

» Do not use sprinklers; instead, water plants at their bases.

» Incorporate the discarded bean stalks and pods into the soil.

» Use 2 gms of carbendazim per kg of seed before planting.

» Use a spray containing either 0.2% chlorothalanil, 0.2% zinc, or 0.1% mancozeb in disease-prone areas.

## Angular leaf spot, *Isariopsis griseola*

### Crop losses

» Reports of yield drops of 10–50% have surfaced.

### Symptoms

» The symptoms appear most obviously on leaves, but they can also be found on petioles, stems, and pods.

» Initially confined to tissue between main veins, lesions on leaves appear as brown patches with a tan or silvery centre and a jagged look.

### Favorable conditions

» This pest is at its most harmful in warm, wet climates.

### Survival and spread

» The fungus spreads from plant to plant through seed and lives on in plant debris.

### Management

» Grow your own healthy plants from good seeds.

» Infected land should be rotated every two years.

**Rust,** *Uromyces appendiculatus* **(syn.** *U. phaseoli***)**

**Losses**

- » Depending on the timing and intensity of infection, yield loss as high as 100%.

**Symptoms**

- » Five to six days after infection, reddish brown, round uredinial pustules appear on the leaves.
- » At 7–9 days after infection, pustules swell significantly and burst, releasing numerous powdery uredospores and possibly some black teliospores as well.
- » The pustules could be as small as a pinhead or as large as 2 or 3 mm in diameter.
- » The undersides of leaves, pods, and sparsely the stems are where rust pustules are most commonly found.
- » If the disease is really bad, the leaves may fall off prematurely.

**Favorable conditions**

- » Climates with high humidity and mild temperatures seem to have the highest rates of the disease.

**Survival and spread**

- » Air currents carry spores (urediniospores) from one area to another.
- » The dark teliospores produced at the conclusion of a growing season remain dormant in the soil throughout the winter, providing a source of inoculum for the following growing season.

**Management**

- » Pick up and dispose of the diseased crop remnants.
- » Rotate non-host crops in every so often.
- » It's important to keep the field weed-free.
- » Diseases can be effectively managed using fungicidal sprays containing 0.2% Mancozeb and 0.05% Bayleton.
- » The 2% foliar spray concentrations of *T. viride* and *T. koningii* were successful in protecting French bean plants against disease.

# White mold, *Sclerotinia sclerotiorum*

## Crop losses

> » There's a chance that the crop will be destroyed entirely.

## Symptoms

> » Beans in the field, in transit, or in storage could lose any and all of their aerial sections.

> » Above-ground plant portions frequently suffer from a watery, mushy decay accompanied by white, fluffy fungal growth.

> » In the white growth and inside the decomposing tissue, small, firm, irregularly shaped sclerotia appear.

> » The pods quickly decay away.

> » White mycelial growth and black sclerotia are extremely indicative of a specific type of mushroom.

## Favorable conditions

> » Crops with a dense canopy in fields with a history of the disease, and seasons with chilly damp conditions during and after flowering, tend to be hit the most by the disease.

## Survival and spread

> » Fungal spores can persist in the soil for more than five years.

> » Wind, tainted irrigation water, and sick seedlings are all potential vectors for disease transmission.

## Management

> » Replace host plants with grains and corn while rotating crops.

> » In order to stop diseases from spreading to nearby secondary hosts, plant in rows perpendicular to the wind.

> » Stay away from fertilisers with high nitrogen content.

> » Create large gaps between rows.

**Web blight,** *Rhizoctonia solani*

## Symptoms

» Sclerotia (asexual) infections manifest as brown necrotic spots (centres) with olive green peripheries, measuring 5-10 mm in diameter.

» The patches grow in size, change shape (becoming more irregular and zonate), and join together.

» Symptoms include the quick development of minute sclerotia and brown mycelium on infected leaves, petioles, flowers, and pods, followed by their demise within 3 to 6 days.

» As the fungus' mycelial growth binds the leaves together, the structure takes on a web-like look.

» Basidiospore spore infections (sexual) manifest as discrete, tiny necrotic, circular lesions of 2–3 mm in diameter.

» Its centres are lighter than the rest of the light brown or brick red exterior.

» Pods that come into touch with the soil are at risk of being attacked by the fungus, which can result in pod blight and the development of a fast transit rot with off- white fungal growth.

## Favorable conditions

» Fungal disease that wreaks havoc in the tropical lowlands of the tropics.

## Survival and spread

» Soil is where the fungus lives out its life cycle as sclerotia or mycelium.

» It can also live on in plant matter and on other host plants after an initial infection.

## Management

» Seeds were treated with a 5.0 g/kg dose of *Trichoderma viride*.

» Neem cake, at a rate of 1 t/ha, is spread on the soil.

» We observed that Carbendazim 0.1% foliar spray was the most effective.

» Cultivars of French beans. Research shows that Arka komal can withstand a lot.

» Mustard cake at 1 t/ha, *T. viride*-enhanced fertilizer-year-old-maize meal at 25 kg/ha, and a 0.1 % solution of carbendazim drenched into the soil all work together to provide a synergistic effect.

**Powdery mildew,** *Erysiphe polygoni*

## Symptoms

»   First appearing as little white spots on the upper sides of older leaves, the disease soon spread to cover the entire leaf, both sides, and the stem.

»   Browning and death of the leaves may occur in extreme situations.

»   It can also infect pods on rare occasions.

»   Although powdery mildew can significantly reduce a plant's growth and yield, it does not typically kill the plant entirely.

»   In warm, damp environments with limited air circulation, this fungus forms fruiting bodies that spread across the plant and take the form of an ashy white powder.

## Favorable conditions

»   In warm, damp circumstances, the fungus spreads swiftly and can completely destroy a bean crop.

## Survival

»   The plants spread by seeds and the soil.

## Management

»   When at all possible, water from below rather than above.

»   If you water in the morning, the sun will dry the foliage and stems as the day progresses.

»   Beans should be given a sturdy structure to climb and lots of room to grow.

»   The fungus is more likely to spread to overcrowded plants.

»   Spores can survive the winter in most places, thus removing infected plant debris before the season ends is crucial.

»   Apply a natural horticultural oil, such Neem or Jojoba, to the plants.

»   Applying 25-30 kg/ha of sulphur as a dust.

»   Use a 0.2% solution of the contact fungicide Karathane.

»   Inhibition of powdery mildew by *Bacillus subtilis*.

**Common bacterial blight,** *Xanthomonas axonopodis* **pv.** *phaseoli*

## Crop losses

» Potential losses might be as high as 45%.

## Symptoms

» Little, water-soaked, light green spots with a broad yellow halo are the hallmark of a leaf lesion's initial appearance.

» Leaves will begin to show dried, brown patches surrounded by a thin, yellow halo.

» Spots could grow and potentially damage leaves if the disease continues to spread. Pods can acquire similarly water-soaked areas that can spread into large, uneven blotches.

» A yellow bacterial crust forms over the affected area during wet weather.

» It's possible for the spot's border to be a ruddy brown, or for the spot itself to be that color.

## Favorable conditions

» The disease spreads more quickly in conditions of high humidity, rain, or both.

» Warm weather facilitates the spread of diseases.

## Survival and spread

» The disease spreads more quickly in conditions of high humidity, rain, or both.

» Warm weather facilitates the spread of diseases.

## Management

» A rigorous seed certification programme ensuring only disease-free seed is used.

» The seed is subjected to a 52°C/20-minute hot water treatment.

» Inoculating seeds with *Rhizobium leguminosarum* bv. *phaseoli*.

» Streptomycin sulphate and sodium hypochlorite treatment of seeds.

» Plant beans every other year at the very least.

» Reduce the number of potential hosts by getting rid of weeds and volunteer beans.

» To prevent water damage, sprinklers should not be placed overhead.

» The use of chemicals like copper hydroxide and potassium methyl dithiocarbamate helps prevent infection in plant foliage.

## Halo blight, *Pseudomonas syringae*

### Symptoms

» The upper side of leaves will reveal small water-soaked areas that have turned necrotic on the underside.

» Around the foci of infection, a halo of yellowish-green tissue develops.

» Leaves can become misshapen if lesions develop on growing leaf tissue.

» Lesions on pods may be a reddish brown colour.

» Lesions on the pods may leak or change colour to a tan.

### Favorable conditions

» When leaves remain damp for long periods of time, disease becomes more serious.

» In temperate climates with enough of inoculum, the disease can really wreak havoc.

### Survival and spread

» The bacterium persists in seeds and crop detritus.

» Natural means of dispersal, such as water splashing and shifting soil, were not taken into account.

### Management

» It's best to sow certified seed cultivated in dry, disease-free climates.

» Sprinkler watering should be avoided since it can supply the necessary moisture and humidity for the development of common blight.

» Common blight-affected fields should be rotated every two to three years, and the infected material should be deep ploughed.

» There are tolerant cultivars out there.

» After harvesting, clean up the field to prevent crop damage.

**Bean yellow mosaic,** *Bean Yellow Mosaic Virus (BYMV)*

## Symptoms

- » Beans, peas, and other leguminous plants are acceptable hosts.
- » Patterns of crinkling, downward cupping, yellow mottling, and mosaic can appear on beans depending on the viral strain, the bean variety, and the time of infection.
- » Foliar symptoms are typically less severe when infection occurs later.
- » Infected pods will have a little deformity and a pale green mottle as they mature.

## Survival and spread

- » This virus spreads through plant seeds.
- » Whiteflies are another vector for this virus.

## Management

- » Put to use resistant cultivars, as BARI Sim-1 and BARI Sim-2.
- » For the best results, only gather seeds from disease-free fields with robust, vigorous plants.
- » It's important to keep the field weed-free.
- » Collecting infected leaves and burning them is necessary.
- » As soon as a few diseased leaves appear on a seedling, the entire plant should be gathered and burned.
- » Spraying insecticides is an option for eliminating vectors. Beginning 20 days after transplanting, spray with Imidacloprid @ 1 mL/L water or Dimethoate @ 2 mL/L water. Repeat every 15 days.
- » Carbofuan, Fensulfothion, Disulfoton, Disyston, and Phorate, applied to the soil at a rate of 1.5 kg active ingredient per hectare, should be used.
- » Disease transmission is being stymied by spraying mineral oil at a concentration of 2%.

**Bean common mosaic,** *Bean Common Mosaic Virus (BCMV)*

## Crop losses

- » Yield losses may vary from 6 to 98% depending on the cultivar and time of infection.

## Symptoms

» Leaves that cup and twist and are a mosaic of bright and dark green.

» Usually, the dark green tissue is bubbling and/or in bands next to the veins.

» Reduced yields can be expected from infected plants due to the development of smaller, curled pods that have an oily look.

» In many cases, this results in a stunting of the plant's development as visible symptoms such as mosaics of light and dark green, leaf roll, malformations, or yellow spots appear.

» Infected young plants may perish from severe vascular necrosis.

» Depending on how far along the plant is when it's infected, it could lose entire sections of itself.

» Vascular necrosis causes a dark discoloration in the pod wall and pod sutures, which can be seen in many pods even in seemingly healthy regions.

## Survival and spread

» Aphids spread the virus from plant to plant, starting with the seeds.

## Management

» It's important to use seeds that have been verified as being free of any diseases.

» Remove diseased plants from the field.

» Eliminating virus transmission through the management of vectors is a priority.

» Aphids could not spread viruses when sprayed with mineral oil at dilutions of 2.5% or 5.0%.

» Utilize beans that are resistant to BCMV, such as the French bean cultivar 'Paulista,' which contains the I gene.

# References

Fening, K. O., Tegbe, R. E., & Adama, I. (2014). On-farm evaluation of homemade pepper extract in the management of pests of cabbage, Brassica oleraceae L., and french beans, Phaseolus vulgaris L., in two agro-ecological zones in Ghana. *African Entomology*, *22*(3), 552-560.

Gogo, E. O., Saidi, M., Ochieng, J. M., Martin, T., Baird, V., & Ngouajio, M. (2014). Microclimate modification and insect pest exclusion using agronet improve pod yield and quality of French bean. *HortScience*, *49*(10), 1298-1304.

Joshi, D., Hooda, K. S., Bhatt, J. C., Mina, B. L., & Gupta, H. S. (2009). Suppressive effects of composts on soil-borne and foliar diseases of French bean in the field in the western Indian Himalayas. *Crop Protection*, *28*(7), 608-615.

Kamaal, N., Akram, M., Pratap, A., & Yadav, P. (2013). Characterization of a new begomovirus and a beta satellite associated with the leaf curl disease of French bean in northern India. *Virus Genes*, *46*, 120-127.

Mageshwaran, V., Mondal, K. K., Kumar, U., & Annapurna, K. (2012). Role of antibiosis on suppression of bacterial common blight disease in French bean by Paenibacillus polymyxa strain HKA-15. *African Journal of Biotechnology*, *11*(60), 12389-12395.

Mondal, A., Shankar, U., Abrol, D. P., Kumar, A., & Singh, A. K. (2018). Incidence of major insect pests on french bean and relation to environmental variables. *Indian Journal of Entomology*, *80*(1), 51-55.

Muvea, A. M., Waiganjo, M. M., Kutima, H. L., Osiemo, Z., Nyasani, J. O., & Subramanian, S. (2014). Attraction of pest thrips (Thysanoptera: Thripidae) infesting French beans to coloured sticky traps with Lurem-TR and its utility for monitoring thrips populations. *International journal of tropical insect science*, *34*(3), 197-206.

Negi, S., Bharat, N. K., & Kumar, M. (2021). Effect of seed biopriming with indigenous PGPR, Rhizobia and Trichoderma sp. on growth, seed yield and incidence of diseases in French bean (Phaseolus vulgaris L.). *Legume Research-An International Journal*, *44*(5), 593-601.

Roy, S. K., Ali, M. S., Mony, F. T. Z., Islam, M. S., & Matin, M. A. (2014). Chemical control of whitefly and aphid insect pest of French bean (Phaseolus vulgaris L.). *Journal of Bioscience and Agriculture Research*, *2*(2), 69-75.

Verma, P., & Gupta, U. P. (2010). Immunological detection of bean common mosaic virus in French bean (Phaseolus vulgaris L.) leaves. *Indian journal of microbiology*, *50*(3), 263-265.

# 14

# Lettuce

## INSECT PESTS

### Green peach aphid, *Myzus persicae*

#### Damage symptoms

» Initial infestations are concentrated on the plant's lower leaves, but as the aphid population increases, it spreads upward and eventually covers the entire plant.

» Withdraw moisture and nutrients from plant tissue, causing young plants to wilt and die.

» Aphids secrete a sticky, sugary substance called honeydew. Black sooty mould develops on honeydew and, while it doesn't hurt the plants per se, it may diminish output by obscuring the sunlight too much.

» Damaged plants are more likely to be infected by aphid-borne viruses, and to succumb to secondary diseases.

» Beet mosaic, beet yellows, and lettuce mosaic are just some of the viral diseasees that green peach aphids can spread.

» Aphid infestations are detrimental to young plants and can taint harvested heads.

#### Pest identification

» Pear-shaped adults with a length between 1.6 and 2.4 mm and a pliable, squishy body. The female, who lacks wings, is a drab yellowish-green colour. The abdomen of the winged migrant form is yellowish green, and it has a dark splotch on its dorsal surface. Cornicles, which resemble tailpipes, can be found on both the male and female versions.

» The nymph resembles the adult in size and shape, but is smaller overall. It has a pale yellow-green colour and has three dark lines across its abdomen.

## Management

» Elimination of farm waste.

» Aphids can be managed with horticultural oil or Neem soap.

» Lady beetles, lacewings, damselflies, flower fly maggots, parasitic wasps, and birds are just some of the natural predators that can be introduced to your garden or encouraged to thrive there.

» Such parasitoids as *Lysiphlebus testaceipes, Aphidius matricariae*, and *Aphelinus semiflavus* prey on this pest.

» When green peach aphid populations are large and relative humidity is high, the fungus *Entomophthora aphidis* may spread and kill off a percentage of the pests.

## Thrips, *Frankliniella occidentalis, Thrips tabaci*

## Damage symptoms

» The feeding of thrips, tiny flying insects, on plants and flowers results in the development of tiny spots, lesions, and a downward curling of the leaves.

» Unless properly managed, their rapid reproduction can result in crop failure in as little as two weeks.

» Large swarms of the small, fast-moving insects can do significant damage to vegetable gardens, and they can be either light green, yellow, or black.

» Thrips create a feeding hole in the epidermis with their single jaw, puncture cells with their maxillary stylets, and drink the sap that leaks out.

» Damage from thrips eating leads slow-growing seedlings to have deformed leaves.

» Brown scarring, giving the leaves a spotted or burned look, is another side effect of feeding.

» The faeces of thrips, which look like tiny black dots, are a telltale sign of damage.

» Young plants lose their leaves if they are severely injured.

## Management

- » Weeds can be managed and intercropping can be done mechanically.
- » Monitor thrip populations using these sticky traps. Thrips are somewhat attracted more to blue traps than yellow ones.
- » Chemical pest control measures.
- » The garlic explodes into a spray of fire.
- » Insects and other predators, like mites and lacewing larva.

## Cut worms, *Agrotis ipsilon, A. segetum*

### Damage symptoms

- » Remove young plants by cutting them at ground level.
- » Several species are known to chew holes in lettuce leaves and even bore into the heads of the lettuce they are eating.
- » The young larva feeds on the underside of the leaf, while the older larva feeds deep within the leaf's head.
- » During the day, cutworms stay in hiding, only emerging at night to feed on plants.

### Pest identification

- » Caterpillars that reach a length of 40-50 mm at maturity can be either grey, brown, or black. They have three sets of legs near the head and five sets of fleshy prolegs.

### Management

- » Cutworms don't do as much destruction where water is consistently provided.
- » Turning over your crops every so often.
- » Parasitic wasps, flies, and ground beetles are some of the common natural enemies.
- » Put out some sprays containing acephate, *Bacillus thuringiensis* sp. aizawai, chlorpyrifos, deltamethrin, diazinon, fenthion, malathion, or trapiomethrin.

## Leaf miner, *Liriomyza* spp.

### Damage symptoms

- » To generate feeding sites and to lay eggs, female adult flies puncture the leaves with their ovipositors.

» The pattern of punctures on the wrapper and cap leaves looks like stippling.

» Mines form at egg-laying punctures as the hatching larvae eat between the upper and lower leaf surfaces.

» Even if the mined leaves are removed during harvest, the lettuce head may still be contaminated by the emerging larvae.

» Caused by the mines, plants are unable to produce as much food, the parts that may be harvested are worthless, and disease can spread.

### Pest identification

» A brilliant yellow triangle patch sits on the upper thorax of an adult, in between the wings.

### Management

» Plants should not have any leaf minor damage.

» Parasitic wasps, such as *Diglyphus isaea, Dacnusa sibirica,* and *Opius pallipes,* are often effective at controlling populations of leaf miners.

» Use a spray containing either Abamectin, Azadirachtin, or Spinosad (Entrust formulation).

» Confidor Energy, Mospilan, Actara, Laser 240 SC, and Decis Mega EW 50 were used for treatment.

» Variety MU06-857 is resistant to leaf miner.

### Whitefly, *Bemisia tabaci*

### Damage symptoms

» Reduced head size, postponed harvest, and leaf chlorosis associated with feeding are just some of the ways that crops can be damaged by whitefly populations that are too large.

» The accumulation of honeydew and sooty mould, as well as contamination caused by the whitefly itself, are equally detrimental to the economy.

» Early fall lettuce plants had been completely wiped out by whiteflies sucking up their phloem sap.

### Pest identification

» The adults of this species are around 1.5 mm in length and have yellowish bodies and white wings.

**Management**

» After harvesting, keep the land clean by getting rid of any agricultural remains that could provide a breeding ground for whiteflies.

» Clear away the weeds that are providing a breeding ground for whiteflies and viruses.

» Avoid whitefly infestation by planting later in the season, as whitefly populations tend to decline between mid-October and November.

» Whitefly and virus populations could be reduced by using sprinklers.

» Mist the area with Imidachloprid or Bifenthrin.

» Parasitizing wasps include several different genera, including the *Encarsia* and *Eretmocerus* families.

» Also, bogeyed bugs, lacewing larvae, and lady beetles feed on whitefly nymphs (*Delphastus pusillus*).

## DISEASES

**Downy mildew, *Bremia lactucae***

**Symptoms**

» Creates irregular, light green to yellow patches on the upper leaf surfaces.

» On the undersides of these patches, a white, fluffy pathogen growth manifests.

» These sores eventually dry up and darken to a brown colour.

» Older leaves are the first to be attacked.

» Leaves with severe infections risk falling victim to the disease.

» Sometimes a pathogen can spread throughout the body, resulting in a darkening of the stem cells.

**Management**

» Remove diseased leaves right away.

» Do not crowd your plants, as this increases the likelihood of disease spreading.

» The disease should be mitigated by the use of drip irrigation, which lessens leaf wetness and humidity.

» Fosetyl-AL, phosphorite, maneb 75 DF, azoxystrobin, or copper hydroxide could be sprayed.

» Varieties like "Avondefiance," "Beatrice," "Musette," and "Valmaine" show some degree of resistance.

## Fusarium wilt, *Fusarium oxysporum* f. sp. *lactucae*

### Symptoms

» Seedlings wither and die.

» An upper taproot red-brown streak is a distinguishing feature of mature plants, and it extends into the crown's cortical tissue.

» The older afflicted crowns have a burnt tip, which may only affect one side of the plant.

» In many cases, vascular tissue in the leaves becomes yellow and develops a brown to black streaking pattern.

» Stunted or unheaded growth is possible in infected plants.

### Management

» The only reliable method of control is to stop growing head lettuce there for a while.

» Romaine lettuce variants are hardier than head lettuce.

## Bottom rot, *Rhizoctonia solani*

### Symptoms

» On the undersides of the leaf petioles and midribs, infected plants develop sunken, reddish-brown lesions of variable depths and diameters.

» Above these wounds, a white to brown mycelium develops.

» It is possible for fungus to spread from leaf to leaf, eventually colonising the entire plant.

### Management

» Do not plant lettuce right after a crop that is vulnerable to *Rhizoctonia*, such alfalfa.

» Effective fungicides include products like Iprodione (Rovral) and Vinclozolin (Ronilan).

» To minimize the spread of disease, fungicide treatment should begin soon after thinning.

**Leaf drop,** *Sclerotinia minor*

**Symptoms**

» The soft, black, and watery decay caused by the pathogen typically affects the main stem or the higher root.

» Infected plants rapidly wilt, collapse, and die due to stem tissue destruction.

» Rotted tissue close to the soil's surface develops dense masses of white fungal growth.

» Fungal structures known as sclerotia are formed on and within rotting host tissue and are hard, black, and irregular in shape.

**Management**

» If you want your lettuce to grow well, you should avoid watering it too often.

» By burying the sclerotia during a deep ploughing, the fungus's infectious spores will die off and the soil will be less conducive to germination.

» Maintain proper drainage and weed-free conditions.

» It has been suggested that growing on a ridge 10 cm high and 15 cm broad, which reduces the amount of time that the plant's lower leaves spend in contact with the soil, could be useful in preventing this disease.

» Fields with problems should implement agricultural rotations with resistant crops, such as maize and grasses.

» Iprodione (Rovral) and vinclozolin (Rovral) are two examples of chemicals that can be used for efficient disease control.

**Bacterial blight,** *Pseudomonas marginalis*

**Symptoms**

» A drenched spot on the outer and more mature leaves.

» As the deterioration spread, the affected tissue progressively darkened to shades of brown, red, or black.

» The affected regions of the leaves were sticky and squishy when it rained, but dried up and became papery and brittle when the weather turned dry.

» The entire plant eventually turned into a smelly, slimy mess.

## Management

» Streptomycin in a spray form may be useful.

» Although the bacterium is likely to survive in crop residue between seasons, crop rotation with non-susceptible crops may be helpful.

» Losses can be minimised by careful growing, harvesting, and packaging as well as speedy pre-cooling.

## Aster yellows (*Phytoplasma*)

### Symptoms

» Young heart leaves turn chlorotic and white.

» The central leaves never fully develop and instead become thicker stubs in the middle of the plant's head.

» Yellowed and curled outer leaves are a sign of plant stress.

» The midribs of damaged leaves develop latex deposits ranging in colour from pink to tan.

### Spread

» Although many different kinds of leafhoppers can spread the phytoplasma, the aster leafhopper (*Macrosteles quadrilineatus*) is the primary vector.

### Management

» Preventing the spread of weeds by removing water sources near lettuce crops.

» Insecticides for eradicating disease spreaders.

» It's best to move new lettuce plants away from older ones in order to prevent the spread of disease.

## References

Barrière, V., Lecompte, F., Nicot, P. C., Maisonneuve, B., Tchamitchian, M., & Lescourret, F. (2014). Lettuce cropping with less pesticides. A review. *Agronomy for sustainable development, 34,* 175-198. McKinney, K. R. (1944). The cabbage looper as a pest of lettuce in the southwest.

Brown, N. A. (1918). *Some bacterial diseases of lettuce.* Department of Agriculture.

Clarkson, J. P., Phelps, K., Whipps, J. M., Young, C. S., Smith, J. A., & Watling, M. (2004). Forecasting Sclerotinia disease on lettuce: toward developing a prediction model for carpogenic germination of sclerotia. *Phytopathology, 94*(3), 268-279.

Hubbard, J. C., & Gerik, J. S. (1993). A new wilt disease of lettuce incited by Fusarium oxysporum f. sp. lactucum forma specialis nov. *Plant Disease*.

Jagger, I. C., & Chandler, N. O. R. M. A. N. (1934). Big vein, a disease of Lettuce. *Phytopathology*, *24*(11).

Oatman, E. R., & Platner, G. R. (1972). An ecological study of lepidopterous pests affecting lettuce in coastal southern California. *Environmental Entomology*, *1*(2), 202-204.

Pink, D. A. C., & KEANE, E. M. (1993). Lettuce: Lactuca sativa L. In *Genetic improvement of vegetable crops* (pp. 543-571). Pergamon.

Santos, C., Monte, J., Vilaça, N., Fonseca, J., Trindade, H., Cortez, I., & Goufo, P. (2021). Evaluation of the potential of agro-industrial waste-based composts to control Botrytis gray mold and soilborne fungal diseases in lettuce. *Processes*, *9*(12), 2231.

Sulvai, F., Chaúque, B. J. M., & Macuvele, D. L. P. (2016). Intercropping of lettuce and onion controls caterpillar thread, Agrotis ípsilon major insect pest of lettuce. *Chemical and Biological Technologies in Agriculture*, *3*(1), 1-5.

# 15 Okra

## INSECT PESTS

### Leaf hopper, *Amrasca biguttula*

#### Damage symptoms

- » This pest assaults the crop when it is still young and vulnerable.
- » The nymphs and adults alike feed on the cell sap.
- » Hence, the leaf margins curl upwards and the entire leaf surface seems scorched.
- » Damaged plants develop abnormally slowly.
- » The plants around here are dying out.
- » Symptomatic of a severe infestation is a leaf's sudden transformation to a dark crimson colour and the appearance of several big necrotic patches.

#### Pest identification

- » The eggs have the appearance of a pear and are long and pale yellow in colour.
- » When frightened, nymphs flee in a diagonal direction.
- » Little (3 mm long), greenish yellow adults with a black patch on the vertex and a black speck on each forewing.

#### Management

- » Hybrid baby corn and okra crop.
- » It is recommended to apply 250 kg/ha of Neem cake to the soil after germination and then repeat the process 30 days later.

- » Spray systemic insecticides such as Acephate 75 SP (1 ml/L), Imidacloprid (0.3 ml/L), or Dimethoate 30 EC 2 ml/L in the early phases of crop development, before flowering. Systemic pesticides are not necessary until fruit harvesting begins, as picking occurs every three to four days.

- » Neem or Pongamia soaps, diluted to 5%, or 4% Neem seed powder extract (NSPE) sprayed on the affected area. A thorough spraying of the underside of the leaves is in need, as this is where the bug is most often to be found.

- » Eggs of *Lymaenon empoascae*, *Anagrus flaveolus*, and *Stethynium triclavatum* are all examples of parasitoids.

- » The mired bug (*Dicyphus hesperus*), the big-eyed insect (*Geocoris* sp.), the lady beetle, the ants, the *Distina albino*, the *Chrysoperla* spp., etc.

- » Okra versus. Jassid population was lower in Arka Anamika, Hisar Unnat, Varsha Uphar, P7, Janardan, EMS-8, Ludhiana Selection-2, and Punjab Padmini.

## Shoot and fruit borer, *Earias vittella* and *E. insulana*

## Damage symptoms

- » Both pre- and post-flowering shoots show signs of infestation.

- » Caterpillars eat the most vulnerable parts of the plant, causing it to wilt and collapse.

- » We eliminate potential growth points.

- » In the evening, they eat the insides of the fruits, which they have bored through.

- » Fruits that are contaminated become unfit for human eating.

## Pest identification

- » The colour of the egg sky is blue.

- » A fully grown caterpillar measures 2 cm in length and is covered in small, stiff bristles and a pattern of black, longitudinal dots.

- » A cocoon's shape resembles a boat.

- » The wingspan of an adult moth is around 2.5 cm, and they feature a small, light longitudinal green line near the middle of the fore wings.

**Favorable conditions**

» Most cases happen after a rainstorm when the air is still damp.

**Management**

» Don't space your plants too far apart.

» It is imperative that the afflicted stems and fruits be gathered and disposed of.

» Pheromone traps will be set up at a rate of 12 per hectare (ha).

» After germination, spread 250 kg per hectare (k/ha) of Neem cake over the area, then do it again two more times at 30-day intervals.

» In a spray every 10 days, use either 1% Neem soap or 4% NSPE.

» Use an efficient contact insecticide, such as Indoxacarb 14.5 SC @ 0.75 ml/L, to spray.

» Inject 10,000 first-instar *Chrysoperla carnea* grubs per hectare.

» The release of *Trichogrmma chilonis* in Tamil Nadu resulted in a 49.22% decrease in the population of the fruit borer (*E. vitella*).

» *E. vitella* damage was reduced after three applications of *B. thuringiensis* (Dipel) at 0.5 kg/ha at weekly intervals.

» Apply *Beauveria bassiana* at 1% WP @ 1500–2000 g in 160–200 L of water/acre.

» *Steinernema feltiae*, an entomopathogenic nematode, can be applied at a rate of 250 billion juveniles per acre.

» Lower borer infestation was found on shoots and fruits of okra cutivars. Selection-2 (Gujarat Agricultural University, Anand), Parbhani Kranti, EMS-8, Parkins Long Green, Karnal special, and Ludhiana Selection-2.

## Aphid, *Aphis gossypii*

**Damage symptoms**

» The colonies of these pests can be easily recognised by the fact that they completely cover the entire shoot, including the buds and undersides of the leaves, with their feeding.

» Juveniles and grownups alike consume the sap.

» Also, they poop honeydew, which has black mould on it.

» Leaves may curl, turn yellow or deformed, develop necrotic areas, shoots may stall and eventually dry out and die from a significant aphid infestation.

## Pest identification

» Aphids have cornicles, which are tubular structures that extend backward from their bodies.

» Not likely to react quickly when startled.

» Adults live in colonies and are tiny, soft-bodied insects that are a greenish brown colour.

» There are two types of adult: those with wings and those without.

## Favorable conditions

» Pest populations tend to flourish when it's dry outside.

» They are more common in younger plants than in older ones.

## Management

» Removal and disposal of diseased stems and other plant pieces.

» Aphids can be deterred from plants by using reflective mulches, such as silver coloured plastic.

» Soap made from Neem or Pongamia (1%), or an extract made from powdered Neem seeds (4% NSPE), should be sprayed liberally.

» Systemic insecticides, such as Dimethoate 30 EC (2 ml/l), Acephate 75 SP, or Acetamiprid, should be sprayed during the pre-flowering stage.

» *Aphidius colemani* is a parasitoid that belongs to the family Cicadidae.

» Green lacewings (*Mallada basalis* and *Chrysoperla carnea*), predatory coccinellids (*Stethorus punctillum*), staphylinid beetles (*Oligota* spp.), the cecidomyiid fly (*Aphidoletis aphidimyza*), and gall midges are only few of the insects that feed on aphids (*Feltiella minuta*).

» The aphid resistance of Pusa A 4 and Gujarat Anand Okra-5.

## Petiole maggot, *Melanagromyza hibisci*

## Damage symptoms

» There is a general trend towards leaf drying.

» When the petioles of such leaves are cut open, the yellow maggots and pupae inside can be seen.

» If pests attack right after germination, it could lead to either death or expansion of the affected area.

» The primary stem may also shatter on occasion.

## Management

» Repeat the application of 250 kg per hectare of Neem cake 30 days after the initial sowing. In the subsequent 10 days following flowering, you should spray your plants with NSPE 4% or Neem soap 1%.

» In the wild, *Eurytoma* sp. parasitizes stem flies at a rate of 36.7%.

## White fly, *Bemesia tabaci*

## Damage symptoms

» The adult and the nymph feed on the sap of leaves.

» Honeydew is secreted, leading to black mould.

» Curling and drying of afflicted leaves is visible.

» As a result, the damaged plants' growth is stunted.

» Veins in the leaf's yellow core form a web that encircles and separates the green tissue islands.

» Whole leaves begin to turn yellow later.

» It's a carrier of the yellow-vein mosaic virus that causes a serious disease.

## Pest identification

» Adults have wings and are between 1.0 and 1.5 mm in length; their bodies are a pale yellow colour with a fine white waxy dusting. Long, conspicuous, and completely white, their hind wings are a defining feature.

## Management

» Plant resistant strains of YVMV.

» Reduce whitefly populations with maize, sorghum, or pearl millet planted as border crops (4 rows).

» Put in two acres' worth of yellow sticky traps for snooping.

» Spray crushed Neem seed powder extract 4% or 1% Neem oil after applying 250 kg/ha of Neem cake during germination and again at 30 DAP.

» At 10-day intervals, spray with Imidacloprid 200 SL at 0.3 ml/L, Thiomethoxam

0.3 g/L, Dimethoate (0.05%), or Metasystox (0.02%). (should not to be sprayed after flowering stage).

» *Encarsia* species, *Eretmocerus* species, and *Chrysocharis pentheus* are all examples of parasitoids.

» Put out 20,000 larvae of the *Chrysoperla carnea* per hectare.

## Fruit worm, *Helicoverpa armigera*

## Damage symptoms

» Leafy greens, young stems, and flower buds are your food sources.

» The caterpillars are initially social and feed together, scraping the chlorophyll content of the leaf lamina to make it look papery white.

» Later on, they skeletonize the leaves, leaving only the veins and petioles visible, and begin cutting irregular holes in the leaves as they eat.

» There has been a massive loss of vegetation.

» Boring, irregularly-holed fruits.

## Pest identification

» Larvae are brown at first, but they eventually turn a greenish colour with darker broken lines running along their sides. An abundance of glowing hairs cover its body. Adults are between 3.7 and 5.0 cms in length.

» The adult moth has a wing span of about 3.7 cms and is brownish or grey in colour with a dark cross bar on the outer margin and dark dots near the costal margins.

## Management

» Hygiene in the field and rogueing.

» The ocimum/basil family is used as a natural insect repellant.

» Adult light traps will be set up at a rate of 1 per acre.

» Putting up birdhouses at a rate of $40/acre.

» For example, planting 100 marigolds per acre to catch the female moths and then harvesting the larvae is a common method.

» The use of pheromone traps, at a rate of four to five per acre, is being implemented.

» Neem oil based WSP with 5% NSKE or Azadirachtin 0.03% (300 ppm) is sprayed on eggs and first instar larva.

» If you see eggs or first-instar larvae at night, spray HaNPV at 250 LE/acre with Jaggery 1 kg, Sandovit 100 ml, or Robin Blue 50 g three times, spaced out by 10-14 days.

» Protect parasitoids like the *Trichogramma chilonis* egg, *Tetrastichus* spp. egg, *Chelonus* spp. egg-larval, and *Telenomus* spp. egg, as well as the Netelia product larva, *Carcelia* spp. larva, *Chaetopthalmus* larva, *Campoletis chlorideae* larva, *Bracon* spp. larva, and others.

» Protect *Chrysoperla carnea*, coccinellids, king corvids, wasps, dragonflies, spiders, robin flies, reduviid bugs, praying mantises, fire ants, and other predators.

» Apply *Beauveria bassiana* at 1% WP @ 1500–2000 g in 160–200 L of water/acre.

» Spread 400-600 gms of *Bacillus thuringiensis* var. gallariae in 200 litres of water per acre.

## Red spider mites, *Tetranychus cinnabarinus*, *Oligonychus coffeae*

### Damage symptoms

» Both nymphs and adults pierce the leaves, and white grey spots can be seen on them.

» Spotted, then brown and falling, affected leaves.

» When mite infestations get bad enough, the plant's canopy is completely coated in webs

» The upper leaf surface develops a characteristic mottling sign.

### Pest identification

» The color of an egg is either white or cream, and it has a round shape.

» Adults range from green to greenish yellow to orange red, and all have two dark dots on their bodies.

### Favorable conditions

» Seasonal warm-ups and dry spells are when mite infestations are most noticeable.

## Management

» Use an acaricide spray, such as Dicofol 18.5 EC at 2.5 ml/L or Wettable Sulfur 50 WP at 3 g/L.

» Neem soap/Pongamia soap 1% can be sprayed as an alternative to synthetic acaricides.

» Good control requires a thorough spraying of the underside of the leaves.

» Anthocorid bugs (*Orius* spp.), green lacewings (*Mallada basalis* and *Chrysoperla carnea*), predatory coccinellids (*Stethorus punctillum*), staphylinid beetles (*Oligota* spp.), cecidomyiid fly (*Anthrocnodax occidentalis*), gall midge (*Feltiella minuta*), etc.

» Spider mites are susceptible to infection by the entomopathogen *Beauveria bassiana*.

# DISEASES

**Damping-off, *Pythium* spp., *Rhizoctonia* spp.**

**Symptoms**

» Damping-off is the premature or immediate demise of newly germinated seeds.

» Poor germination occurs when an infection exists prior to the emergence of the seedling.

» "Damping-off" occurs when degradation occurs after seedlings have emerged, causing them to collapse or perish.

» Emerging seedlings get a sore where their delicate stem first makes touch with the earth.

» Softening of the tissues beneath the lesion causes the seedlings to collapse.

**Favorable conditions**

» Diseases thrive in conditions that are conducive to infection and development, such as high humidity, high soil moisture, cloudy weather, and temperatures below 24°C for a few days.

» Congested seedlings, heavy rainfall, poor drainage, and an abundance of soil solutes all contribute to damping-off caused by pathogens.

## Survival and spread

» Major vectors of transmission are the ground, plant seeds, and water.

» Conidia can be dispersed secondarily by wind or rain.

## Management

» When all agricultural leftovers are totally digested beforehand, the occurrence of "damping-off" diseasees decreases.

» To keep humidity levels around plants low, avoid watering them too frequently.

» *Trichoderma viride* (3-5 g/kg of seed) or Thiram (2-4 g/kg of seed) seed treatment.

» Dithane M-45 (0.2%) or Bavistin (0.1%) soil drenching.

## Leaf spot, *Cercospora abelmoschi*

## Symptoms

» The fungus begins as a little spot on the underside of the leaf, but it quickly spreads and eventually covers the entire leaf.

» Produces splotchy black areas with sharp edges.

» Photosynthetic area is drastically reduced due to a pathogen.

» After drooping and rolling, leaves fall to the ground rapidly.

## Favorable conditions

» Normal throughout the rainy season.

## Spread and survival

» Plants infected with the fungus will continue to live.

» Conidia and stromata are the fungal reproductive structures responsible for the fungi's survival on crop waste in soil.

» The wind carries the conidia to new locations.

## Management

» Collecting and burning defoliated leaves in the field can help reduce inoculum levels directly where they are produced.

» Effective management of leaf spot disease requires three sprayings with

0.05% Bitertanol or 0.025% Tridimefon or 0.8% Bordeaux mixture at 500 L/ ha beginning at the onset of symptoms and continuing at weekly intervals.

» Use a 0.25 % Mancozeb spray.

## Powdery mildew, *Erysiphe cichoracearum*

### Symptoms

» Tiny, white to dirty-grey patches often form on the skin as a symptom of disease.

» Greyish powdery growth can be seen on both the underside and the top of the leaf and the fruit.

» As they grow, they take on a powdered texture.

» The green surface of the host could be completely buried behind a powdery coating.

» It is possible that patches will join together to cover the entire plant.

» The leaves of a severely diseased plant may curl upward and get a burnt appearance.

» Defoliation and early mortality of plant parts are the results of a severe infection.

» Reduction in fruit production caused by this factor is significant.

### Favorable conditions

» Conditions favourable to the spread of disease include high humidity during the day and low humidity at night, which leads to the creation of dew.

» A dry climate with temperatures between 15 and 30 °C, especially between the months of September and December, is conducive to the spread of disease.

### Survival

» Holds out in contaminated crop waste or a different host.

» By making slumbering structures like mycelium or cleistothecia.

### Management

» Apply water from above (washes fungus from leaves and reduces viability).

» Sow seeds as soon as feasible.

» For optimal relative humidity reduction, keep plant spacing at its optimal level.

» The combination of 4% NSKE and 1% Neem oil as a spray was very successful in preventing the spread of disease.

» Three or four sprayings, each 15 days apart, using inorganic Sulfur 0.25 % or Dinocap 0.1 %.

### Fusarium wilt, *Fusarium oxysporum* f. sp. *vasinfectum*

### Symptoms

» The leaves begin to yellow and wilt before falling off.

» Often, the lower leaves are the first to go.

» Brown staining in the ring just beneath the bark is common when a stem or main root is cut across.

» Plant wilting is typically slow, but might be more noticeable after a soaking summer rain.

» The growth of infected plants is impeded if the infection is severe.

### Favorable conditions

» Warm weather facilitates the spread of disease.

### Spread and survival

» This fungus originates in the ground.

» Infected seed, polluted machinery, or unclean hands can all bring fungus to a field.

### Management

» In order to avoid disease, only plant certified seed.

» Use 3 gms of Mancozeb per kg of seed for seed treatment.

» Use cereals and grasses in rotation if at all practicable.

» Don't plant in soil that has a history of *Fusarium* wilt.

» When possible, plough fields deeply and let them lie fallow for two to three months.

» Use of lime or agricultural manure can increase the soil pH.

» Copper oxy- chloride at 0.25% should be sprayed all over the field.

» Get rid of the root-knot nematodes.

## Blossom and fruit blight, *Choanephora cucurbitarum*

### Symptoms

» The okra blossom, or the blossom end of young fruit, is the first target of the fungus' invasion.

» The flowers and the tips of the fruits develop a fuzzy growth that resembles bread mould.

» Fruit ripens softer than usual, turns brown, and may grow noticeably longer.

» In time, the mould could spread and cover the fruit entirely.

» A plant's lower fruits are more vulnerable to contamination.

### Favorable conditions

» Where there is a lot of heat and moisture, it thrives and causes the biggest problems.

» When the right conditions of warmth and dampness are met, fungi flourish.

### Survival

» The soil is a reservoir for *Choanephora cucurbitarum*

### Management

» To combat the fungus's love of damp environments, gardening best practises include planting on raised beds, increasing plant spacing, and using foggers.

» Avoid getting the leaves wet by watering the plant from below, and water first thing in the morning to promote evaporation.

## Enation leaf curl virus

### Symptoms

» Deformations include thickening and curving of the leaves along their adaxial axis, as well as the development of mild or bold enations on the underside of the leaves.

» Petiole, main stem, and branch twisting.

» Infected leaves develop a leathery thickness.

» Plant development is slowed down.

» Infected plants produce little, misshapen fruit that can't be sold.

## Spread

» White flies spread the infection.

## Management

» Track the whitefly population with the help of yellow sticky traps.

» The removal of infected plants from the field is a need.

» Mineral oil (2%), followed by either 0.05% dimethoate, 0.02% metasystox, or 0.05% monocrotophos sprayed on the leaves.

» Applying Furadon to the soil at a rate of 1.5 kg a.i. /ha just before planting seeds is also highly recommended.

## Yellow vein mosaic disease, *Yellow Vein Mosaic Virus* (YVMV)

### Crop losses

» If the plants are infected within the first 20 days after germination, the yield loss can reach as high as 98%. There was an estimated 83% and 49% yield loss in plants that were infected either 35 or 50 days after germination.

## Symptoms

» Symptomatic veins and veinlets are yellow, with the intervenous tissue remaining green.

» If the infection is bad enough, the younger leaves will turn yellow, shrink in size, and the plant will become severely stunted.

» In the early stages of development, fruits turn yellow and harden.

» Diseased plants are stunted and produce only a few yellow fruits.

» Infected plants are less likely to bloom, and their fruits may be smaller and tougher if they do develop.

» Fruits from infected plants turn out yellow or white and are unsellable.

## Favorable conditions

» More whiteflies and more cases of the virus were recorded between March and June, when the weather is warm and dry and the vector population is high.

## Survival and spread

» Plants like *Hibiscus tetraphyllus*, *Croton sparsiflora*, and *Ageratum* spp. that have been infected by the virus serve as collateral hosts, where the virus can persist for a long time.

> Whiteflies, *Bemisia tabaci*, are responsible for spreading the Gemini virus that causes YVMV.

## Management

> Plant certified seeds that haven't been contaminated with any diseases.

> The immediate removal of any infected plants from the garden.

> At least 60 days before planting okra, sow four to five rows of sorghum, pearl millet, or maize around the perimeter of the field.

> When combined with border cropping and three to four foliar sprays of either 0.1% Dimethoate or Monocrotophos at 10-day intervals, these methods have proven to be more effective than either treatment alone.

> With a litre of water, use 2.5 ml of Chlorpyriphos and 2 ml of Neem oil.

> Choose okra varieties with higher disease resistance, such as Arka Anamika, Arka Abhay, Pusa Sawani, Harbajan, Parbhani Kranti, Janardhan, Haritha, Hisar Unnat, Hisar Naveen, Azad Bhindi -1, Azad Bhindi -3, and Varsha Uphar.

# References

Anitha, K. R., & Nandihalli, B. S. (2008). Seasonal incidence of sucking pests in okra ecosystem. *Karnataka journal of Agricultural sciences*, *21*(1), 137-138.

Arain, A. R., Jiskani, M. M., Wagan, K. H., Khuhro, S. N., & Khaskheli, M. I. (2012). Incidence and chemical control of okra leaf spot disease. *Pak. J. Bot*, *44*(5), 1769-1774.

Afzal, S. A. I. M. A., Tariq, S., Sultana, V., Ara, J., & Ehteshamul-Haque, S. (2013). Managing the root diseases of okra with endo-root plant growth promoting Pseudomonas and Trichoderma viride associated with healthy okra roots. *Pak. J. Bot*, *45*(4), 1455-1460.

Ali, S., Khan, M. A., Habib, A., Rasheed, S., & Iftikhar, Y. (2005). Management of yellow vein mosaic disease of okra through pesticide/bio-pesticide and suitable cultivars. *International journal of agriculture and biology*, *7*(1), 145-147.

Dhamdhere, S. V., Bahadur, J., & Misra, U. S. (1984). Studies on occurrence and succession of pests of okra at Gwalior. *Indian Journal of plant protection*, *12*(1), 9-12.

Gupta, S., Sharma, R. K., Gupta, R. K., Sinha, S. R., Singh, R., & Gajbhiye, V. T. (2009). Persistence of new insecticides and their efficacy against insect pests of okra. *Bulletin of environmental contamination and toxicology*, *82*, 243-247.

Kumar, A., Kumar, P., & Nadendla, R. (2013). A review on: Abelmoschus esculentus (Okra). *International Research Journal of Pharmaceutical and Applied Sciences, 3*(4), 129-132.

Mandal, S. K., Sah, S. B., & Gupta, S. C. (2007). Management of insect pests on okra with biopesticides and chemicals. *Annals of Plant Protection Sciences, 15*(1), 87-91.

Ndunguru, J., & Rajabu, A. C. (2004). Effect of okra mosaic virus disease on the above-ground morphological yield components of okra in Tanzania. *Scientia Horticulturae, 99*(3-4), 225-235.

Obeng-Ofori, D., & Sackey, J. (2003). Field evaluation of non-synthetic insecticides for the management of insect pests of okra Abelmoschus esculentus (L.) Moench in Ghana. *SINET: Ethiopian Journal of Science, 26*(2), 145-150.

# 16

# Onion and Garlic

## INSECT PESTS

### Thrips, *Thrips tabaci*

**Damage symptoms**

- » Between the leaf sheaths and the stems, you can see a swarm of nymphs and adults lacerating the leaf epidermis and sucking the leaking cell sap.

- » Spots of silvery colour appear on the damaged leaves, which eventually become brown.

- » The plant wilts and dies as the leaves become twisted from the tips down.

- » Heavy infestation results in the death of seedlings and stunted plant growth.

- » Thrips reduce both bulb size and yield.

**Pest identification**

- » Eggs range in colour from white to yellow, have the shape of a kidney bean, and are quite little.

- » The first two instars of a larva's development have a body colour range from white to pale yellow and are long and slender. Look like adults, except they can't fly. The eyes are a deep blue, and the antennae are relatively short.

- » Both the pre-pupa and the pupa stages are a muted yellow to a light brown colour, and their bodies are much stockier than those of earlier instars. The antennae are tucked into the head and the beginnings of wings can be seen.

- » Long, yellowish brown body with two sets of fringed (hairy) wings; adult size is around 1.5 mm. Beak-like features of the mouth and 7-segmented antennae are characteristic.

## Management

- » Ten per hectare, use yellow sticky traps.

- » Plant two rows of corn as a protective barrier around your onion crop.

- » Neem seed powder extract 4% or Neem soap 1% sprayed every 10 days after planting, and Neem cake applied at 250 kg/ha to the soil 30 and 60 days after planting.

- » Granular phorate or carbofuran (1 kg a.i. /ha) applied to the soil.

- » Alternating sprays of an Azadirachtin formulation (more than 10,000 ppm) at 2 ml/L with those of Fipronil 5 SC at 1 ml/L, Lamda cyhalothrin 5 EC at 0.5 ml/L, and Acephate 75 SP at 1 g/L.

- » Set predators such as *Scymnus nubilis*, *Chrysoperla* spp., and *Latus* spp. cutivars onion Some resistant/tolerant varieties include Arka Lalima, White Persian, Grano, Sweet Spanish, Crystal Wax, Spanish White, Bombay White, Pusa Red, Udaipur 103, N 780, N 53, and White Poona Red.

## Maggot, *Delia antiqua*

### Damage symptoms

- » The larva alone is harmful, as it enters the plant from its base utilising its hooked mouth parts.

- » Once it has fed for 14-21 days, it will bore into the bulb by cutting through the subterranean stem.

- » Seedlings that are injured first wilt, then become floppy and eventually die.

- » Seedlings that are attacked by maggots often perish before the maggots can fully mature, so they move on to nearby plants.

- » When maggots of the second-generation feast on emerging bulbs, the consequence is often deformed growth and tissue decay.

- » Third generation maggots cause late-season onion bulbs to be ruined and unsellable.

- » Infected vegetation turns a drab brown colour and eventually dies.

### Pest identification

- » Eggs: 1.25 mm long, white, elongated eggs.

- » Maggots are roughly 8 mm in length, have no legs, and are a creamy white colour.

- » Pupae are a length of 7 mm and a chestnut colour.
- » Adults of this species are distinguishable from common house flies by their smaller size, longer legs, and thinner build; they also rest with their wings overlapping.

## Management

- » In early spring, before the adult develops, infected bulbs should be buried in pits and covered with about 30 cm of soil.
- » Don't follow up onion crops with more onions.
- » To take advantage of the warmer soil conditions, planting should take place in the fall.
- » Make use of sticky yellow traps.
- » Before planting, the soil should be treated with phospate and the crops should be rotated regularly.
- » *Coenosia tigrina* and *Scatophaga stercoraria* are predatory flies; the wasp *Aphaerata pallipes*; the beetle *Aleochara bilineata*; and the fungus *Entomopthora muscae* are all natural enemies.

## Armyworm, *Spodoptera exigua*

## Damage symptoms

- » Caterpillars devour fields and nurseries in search of leaves to munch on in a communal feeding frenzy.
- » As a result of their feeding, fewer bulbs will develop.

## Pest identification

- » Grayish tones characterise these moths.

## Management

- » To protect against pests, plant sorghum or maize in four rows all the way around your onion/garlic crop.
- » Alternate growing onions with a crop of a non-host cereal, cucurbits, or crucifers.
- » You can attract ovipositing insects to your garden by growing garlic, Ocimum/ Basil, and marigold.
- » Set up bird perches at a rate of 50 per hectare to attract predatory birds like crows, mynahs, and kingfishers.

» Set out 4-5 pheromone traps per acre to track the number of flying moth adults.

» Every two to three weeks, you should switch out the bait.

» Light trap installation at a rate of 1 per acre.

» From the time flowers begin to form, you should inundate your land four to five times with 40,000 *Tricogramma* spp. or *T. pretiosum*.

» Protect parasitoids like the *Campylobacter* sp. larva, the *Tetrastichus* sp. egg, the *Telenomus* sp. egg, etc.

» Protect predators like the *Chrysoperla zastrowi sillemi*, coccinellids, King crow, common mynah, wasp, dragonfly, spider, robber fly, reduviid bug, praying mantis, fire ants, big eyed bugs (*Geocoris* sp), pentatomid insect (*Eocanthecona furcellata*), earwigs, ground beetles, rove

» Use 50–300 million *Steinernema feltiae* juveniles/hectare of entomopathogenic nematodes (EPNs).

## Leek moth, *Acrolepiopsis assectella*

### Damage symptoms

» The damage caused by the larvae's tunnelling into the leaves makes the plants more susceptible to bacterial and fungal infections.

» The larvae eat both the outer and inner layers of the leaves.

» As they worked their way into the plant's core, they pierced the folded leaves along the way, leaving behind a trail of tiny holes.

» In the adult plant, the longitudinal tunnels made by the larvae in the middle leaves have faded to a dull grey colour.

### Pest identification

» The wingspan of an adult is about 8 mm. The upper wings are an ashy grey that almost looks black. There is a big white triangle on the forewing, and there are several smaller white patterns on the wing itself. The back wings are a dark grey that borders on black.

» The full-grown larvae are just about 7–8 mm long.

» The cocoon has the appearance of a white, loose net.

## Management

» When applied at the right time, insecticides, including organic formulations, can reduce damage caused by leek moths and keep populations under control.

» Rotating crops.

» Planting was put off.

» Getting rid of dead and infected leaves.

» Eliminating eggs or larvae.

» An early harvest (to avoid damage by last generation larvae and population build-up).

» Keeping vulnerable crops in areas free of pests.

» In the aftermath of harvest, all plant remnants must be eliminated.

» Protecting leeks from predators and the elements by netting them up before they bloom and removing their outer leaves before winter leaves appear is an effective strategy.

» Floating, lightweight row coverings can prevent leek moth damage to young plants.

## Leaf miners, *Lyriomyza* spp.

## Damage symptoms

» The cotyledons of seeds can be damaged by adult leaf miners, causing the seedlings to develop slowly.

» In order to feed, larvae mine between the upper and lower leaf surfaces, leaving behind a maze of thin, white tunnels that eventually deepen as the larvae develop.

» Mining leaves excessively lowers their photosynthetic capacity, makes them unsellable, and makes them an easy target for pathogens.

» White, twisting, and very fine trails on the leaves.

» White spots on leaves and early leaf drop are two symptoms of leaf injury caused by heavy mining.

» Yield loss due to early infection is possible.

## Pest identification

» Little, black or grey flies with yellow markings represent adults. Bristles that are long and rigid cover the body.

» The full-grown length of a larva is around 0.25 inches, and it is an almost transparent white or yellow.

## Management

» Before planting, inspect the transplants for leaf minor damage.

» Immediately after harvesting, pull plants out of the ground.

» Insecticides like Abamectin (Agri-mek 0.15) should be sprayed at a rate of 0.009 to 0.019 lb ai/acre, Cypermethrin (Holster) at a rate of 0.04 to 0.1 lb a.i. /acre, Lambda-cyhalothrin (Warrior II) at a rate of 0.015 to 0.025 lb Spraying them too often will also kill off too many of their natural predators.

» Control is best achieved with the help of parasitic wasps.

» Parasitoid, *Diglyphus begini*, has been released.

## Lesser bulb fly, *Eumerus* spp.

## Damage symptoms

» Only onions grown from sets, bulbs, or multipliers are vulnerable.

» Larvae devour the tissue around the bulb, leading to decay and the wilting or loss of the plant's leaves.

» Particularly susceptible are plants that are already under stress or suffering from diseases or decay.

» When onion bulb fly larvae are present, only one brown, withered leaf will sprout from the bulb.

» Infested bulbs are mushy to the touch after harvest and quickly deteriorate into a bulb of rot when they are stored.

» After harvesting, remove any bulbs that look like they might have been infected.

## Pest identification

» Adults of this species are around 8 mm in length, black-green in colour, and marked with a few white lunate patterns on an otherwise nearly hairless abdomen.

» In appearance, the legless larvae are a wrinkly gray-yellow. Larvae can grow to a maximum length of around 1.3 cm.

## Survival and spread

» Larvae overwinter in soil to depths of 8 cms, but most often in contaminated onion bulbs that are left in the field.

## Management

» Before planting, dormant bulbs can be heated in water at 40 °C for an hour.

» At the first sign of these onion pests in the bulbs, pull up and remove any affected plants before they can return to the soil and reproduce. Onion pests tend to attack in clusters, or one bulb at a time, down a row. This makes it simpler to identify plants in trouble and remove them.

» Spreading onion seedlings across the garden and digging them out can assist.

» To prevent the onion fly from laying its eggs in the soil, make sure to press down the dirt around the seedlings and cover them with paper, cloth, or any insect-proof material.

» Adult onion flies can be discouraged from laying their eggs at the base of the plants by spreading a fine layer of sand or wood ash around the plants.

## Bulb mite, *Rhizoglyphus* spp., *Tyrophagus* spp.

## Crop losses

» Have a devastating impact on garlic harvests worldwide, lowering yields by between 23 and 32 %.

## Damage symptoms

» Damage from mite attacks manifested as twisted, curled leaves that did not unfold normally, establishing a microclimate on the upper leaf surface that was ideal for the colonisation of all mite life stages along the midrib.

» Most diseased leaves drooped at an angle, with their tips tucked under the adjacent set of younger leaves.

» Severe infestations caused noticeable yellowing of the leaves, most noticeably around the leaf's midrib and its border.

» As the mite infests a leaf, it usually does so along the midrib, where it can do the most harm.

» Infected plants have stunted development and leaves that are coiled and twisted in a specific way.

» Stored garlic bulbs were ravaged by the mite, which led to their ultimate drying up or deterioration because of surface damage.

## Pest identification

» Microscopic mites are transparent white in colour and have a cigar form that tapers from the head to the tail.

» They stand on only four legs, all of which are positioned close to the head.

» The length of an adult mite is around 200 m while the width is between 36 and 52 m.

» Eggs have thin silk webbing to hold them in place.

## Management

» Using an acaricide on garlic is as easy as dipping the bulbs and letting them dry in the sun for 15 days.

» While planting garlic, make sure you use only pristine seed cloves.

» Crops that are not as easily damaged by bulb mites should be rotated in with onions and garlic.

» Try soaking garlic cloves in hot water before planting.

» Protect predators like lacewings (*Mallada basalis* and *Chrysoperla zastrowi sillemi*), mites (*Amblyseius alstoniae, A. womersleyi, A. fallacies*, and *Phytoseiulus persimilis*), coccinellids (*Stethorus punctillum*), staphylinid beetles (*Oligota* spp.

» In terms of garlic, the most resistant varieties were identified as Katki and G-50.

## Red spider mite, *Tetranychus urticae*

## Damage symptoms

» The adults and their young cause damage by feeding on the undersides of leaves, which causes the leaves to curl, change colour, and become stunted.

» The feeding punctures seem like tiny spots scattered across the upper leaf surface.

» The mites like to make their homes around the plant's veins and midrib.

» Typically, the silk webbing that these mites make is obvious.

» The leaves turn yellow and may fall off as a result of the bleaching.

» They also cause damage to stored inflorescence and dormant bulbs.

» Mites eat the epidermal cells of bulb scales, killing them and causing the scales to fall off.

## Pest identification

» Eggs are tiny spheres that are about 1/254 of an inch in diameter. They are smooth, glossy, and the colour of straw.

» The larva is somewhat bigger than the egg, has a rosy color, and can move around on three sets of legs.

» The nymph goes through two distinct developmental stages, the protonymph and the deutonymph. The nymphal stage can be distinguished from the larval stage because it is larger, changes colour to a reddish or greenish tone, and has an extra pair of legs.

» A mature female measures about 1/50 of an inch in length and is reddish brown with an oval shape.

» A black mark on one side of their otherwise colourless bodies distinguishes males, who are also smaller and more wedge-shaped.

## Management

» Neem cake applied to the soil 30 and 60 days after planting, followed by sprays of Neem seed powder extract 4% or Neem soap 1% every 10 days.

» Apply acaricides, such as Wettable Sulfur 80 WP at a concentration of 3 g/L.

» Protect predators like lacewings (*Mallada basalis* and *Chrysoperla zastrowi sillemi*), mites (*Amblyseius alstoniae, A. womersleyi, A. fallacies,* and *Phytoseiulus persimilis*), coccinellids (*Stethorus punctillum*), staphylinid beetles (*Oligota* spp.)

# DISEASES

## Basal rot, *Fusarium oxysporum* f. sp. *cepae*

## Symptoms

» Yellowing of leaves and tip dieback are the first visible symptoms in the field.

» As the disease worsens, the plant as a whole may collapse, and if it is plucked up, it may do so without any roots attached.

» Pinkish brown discoloration of the onion's base plate and secondary bacterial rots are possible outcomes of this condition.

» Infections late in the season can delay the appearance of symptoms until the onions are already in storage.

## Favorable conditions

» When soil temperatures are high (about 29 °C), basal rot is common.

## Survival and spread

» The virus was found in the soil.

» Infection can enter at the root directly or through a wound.

## Management

» The storage area needs proper ventilation to prevent the buildup of harmful gases.

» Lower symptoms are experienced while storing at room temperature (between 8 and 15 °C). Tissues degrade and dry very quickly at 30°C.

» Fields that have been growing onions for multiple years without proper crop rotation are particularly susceptible to Fusarium basal rot.

» Harvesting and handling should be done carefully to prevent mechanical injuries, and curing is advised.

» Seedlings can be protected from pathogens by dipping them in a solution containing antagonists such as *Pseudomonas cepacia* and *Trichoderma viride*.

» Using a polythene sheet of 750 gauze to solarize the soil during the summer for 20-30 days minimises disease prevalence.

» Good control of basal rot in seed crops can be achieved with pre-planting soil application of *T. viride* at a rate of 1250 g + 50 kg FYM. cutivars onion It has been stated that Arka Lalima, Arka Niketan, and Arka Pragati are resistant or tolerant to basal rot.

## Purple blotch, *Alternaria porri*

## Symptoms

» Lesions filled with water and white in the middle are the first sign of trouble.

» The lesions' borders darken to brown or purple, while the surrounding areas of the leaf, both above and below, turn yellow.

» Once some time has passed, concentric rings ranging in colour from dark brown to black develop across the lesions.

» The fungus is sporulating in these spots.

» Lesions can girdle a leaf as the disease proceeds, causing it to wilt and eventually die.

» Similar signs, including the potential for the seed stalk to collapse and the subsequent stunting of the seed, are seen in infected stalks.

» Most cases of bulb infection involve entry at the neck.

» In the event of a fungal invasion, the affected portion of the bulb will first appear bright yellow, before settling into a distinctive red wine color.

## Management

» Planting onion bulbs that have been preheated at 35 °C for 8 hours.

» Plants that were planted in the summer were protected from the disease.

» Using surface watering instead of spray irrigation, proper field drainage, and appropriate plant spacing can all help to minimise leaf wetness and, by extension, disease development.

» Mancozeb spraying at a concentration of 0.2 %. cutivars onion Tolerant varieties include the Arkas: Kalyan, Kirtiman, Lalima, and Dark Red from Agrofound.

» Mixing coarse cereals into your crop rotation.

## Downy mildew, *Peronospora destructor*

## Symptoms

» Infections in the field typically start in localised areas before spreading rapidly.

» The pathogen's brownish-purple velvet-like sporulation on otherwise healthy green leaves is often the first sign of the disease.

» Lesions, initially smaller and lighter in colour than the rest of the leaf, can grow to encircle the entire leaf as the disease spreads.

» Collapse of the leaf tissue occurs when these lesions advance from a pale yellow to a brown necrosis.

## Survival and spread

» Although mycelium has been seen on and within seeds, real seeds do not aid in the transmission of the fungus from one growing season to the next. The fungus destroys the seed stalks in a seed crop.

» In many regions, infected bulbs are used to propagate the crop, and oospores can be found in infected crop wastes.

» When infected bulbs are planted, the fungus travels up the stems and leaves to disperse its sickness.

## Management

» To prevent the spread of the fungus, onion sets that have been affected should not be planted.

» Clean up the area by getting rid of any dead plants and removing any mounds that have accumulated.

» If you want to save water and time, utilise furrow irrigation instead of sprinklers and plan your planting rows to take advantage of the wind.

» Infected areas can benefit from a 3- to 4-year break from growing onions.

» To prevent the spread of downy mildew, it is advised to apply a fungicidal spray such as Difolatan, Cumin, or Ridomil at a concentration of 0.1% on a routine basis.

## Smut, *Uromyces cepulae*

## Symptoms

» Within the first six weeks of sprouting, infected seedlings generally perish.

» The cotyledons' dark spots are the first features visible once the seedlings have emerged from the soil.

» Large lesions cause leaves to droop downward and form blisters on older plants that have been raised.

» Foliage, leaf sheaths, and bulbs may all develop streaks.

» A dark, powdery mass of spores can be found within mature tumours.

» As the infection spreads from leaf to leaf, the affected plants' growth is inhibited.

## Survival and spread

» Spore balls allow the fungus to survive for up to 15 years in infested soil.

» It can live as a saprophyte for long periods of time in the ground.

» Both onion bulbs and transplants contain the fungus, making them key vectors in the disease's rapid spread.

» Tools can also be used to aid in the dissemination.

» Local dispersal can also occur via wind-borne soil and surface drainage water.

## Management

» Planting healthy onion sets or transplants into infected soil does not guarantee that they will become infected.

» The disease can be prevented in onion crops by using transplants.

» Disease can also be reduced by not growing onions for three years or more.

» Soil-borne disease can be prevented by treating seeds with Thiram or Captan at a rate of 3 g/kg of seed.

» Before planting, you can prepare the soil by applying 0.2% Thiram and 0.2% Captan.

## Pink root, *Phoma (Pyrenochaeta) terrestris*

## Symptoms

» It doesn't matter how old an onion is, the fungus will attack it.

» Light pink appears on infected roots, which then darken to a deeper pink or red and lastly a purple-brown as they shrivel and die.

» The infected roots, which turn black and die off, are the source of the problem.

» Even if new roots emerge, the fungus may eventually kill them.

» Infected plants may show symptoms of nutrient deficiency or drought, with leaves turning white, yellow, or brown from the tips and ultimately dying.

» There are fewer leaves and they're smaller, and the plants are more vulnerable to being uprooted.

» Early-season infections cause plants to develop bulbs too soon and cause more harm than later infections.

» Infected plants typically produce smaller than average bulbs that have less overall value.

» Older plants usually survive, but bulb formation is stunted and harvests are meagre.

» Insects don't eat bulbs, but they can eat through the protective outer scales.

## Favorable conditions

» Diseases flourish at temperatures between 24 and 28 °C in the soil.

## Survival and spread

» Mycelia, sick roots, and agricultural detritus are all places where fungi can live in the soil and continue to spread for years.

» The fungus is mobile, and it can be disseminated by water used for irrigation or through the soil itself.

» Most cases of infection are caused by mycelia in the soil, which means that this pathogen is spread through the soil.

» Onion sets are a common vector for infection.

## Management

» Losses from this disease can be kept to a minimum by delaying planting until the soil is too warm to support disease development.

» Losses can be minimised by rotating to non-host crops like cereals for an extended period of time (four to six years).

» Pink root can be reduced and the yield of saleable bulbs increased with the use of soil solarization or fumigation.

## Rust, *Puccinia allii*

## Symptoms

» Spots or specks of white to yellow colour appear first on leaves, progressing to yellow and orange as the fungus progresses through its life cycle within the leaf tissue.

» Uredinia are pustules that develop from yellowish-orange patches and are home to infectious urediniospores.

» Black telia with dark teliospores grow from black pustules later on.

» When leaves are severely diseased, they might take on an orange color.

» Infected plants cannot survive for long.

**Favorable conditions**

» Humidity and a lack of precipitation are ideal conditions for the spread of a disease that attacks garlic plants and is carried in the air.

**Management**

» It is recommended that beginning in February, sulphate of potash be hoed into the soil at increasing concentrations around the plants.

» Spraying with sulphur compounds or Dithane should be done routinely.

» Application of a fungicide to the foliage before a disease appears is essential.

**White rot, *Sclerotium cepivorum***

**Symptoms**

» Older leaves usually turn yellow, wilt, and fall off as the first signs of disease.

» The rot caused by the fungus' invasion of the root system and basal plate ultimately leads to the collapse of the plant's foliage.

» A white mycelial growth and a mushy rot emerge slowly at the bulb's base.

» The affected tissues get covered in many sclerotia.

» Clusters of plants in the field are a common target for this disease.

» But, when the fungus is plentiful in the soil and conditions are suitable for disease growth, vast groups of plants may perish abruptly.

**Favorable conditions**

» In the case of winter onions, the disease is at its worst during warm spells in the fall or spring.

**Survival and spread**

» The sclerotia can live in the soil for up to eight years.

» Primary inoculum is made up of tiny black sclerotia that have accumulated in the cracks of Allium spp. plants over the course of several years.

» The sclerotia are spread by the flood water from one field to another.

**Management**

» Utilize only disease-free seedlings, and don't water or fertilise the field with anything that could spread pests.

» Onions and garlic are planted early.

» To solarize soil, place a clear sheet on it and wait 8-11 weeks during the summer.

» There is evidence that sclerotial populations in soil can be reduced through methods such as flooding and the application of natural and synthetic sclerotia germination stimulants, both of which may help mitigate losses caused by this disease.

» Infected plants must be uprooted and destroyed.

» Solarized soil is treated with *Trichoderma harzianum*, *Trichoderma viride*, *Trichoderma virens*, and *Bacillus subtilis*.

## Blue mold rot, *Penicillium* spp.

### Symptoms

» The earliest signs are soggy, pale yellow sores.

» Quickly, the damaged areas get blanketed in distinctive blue-green spores.

» When damaged bulbs are sliced open, the fleshy scales may be soaked in water and appear a pale tan or grey.

» Bulbs can become waterlogged, mushy, and tough as deterioration progresses.

» There is typically a musty odour.

### Favorable conditions

» Temperatures between 21 and 25 °C with considerable humidity are ideal.

### Survival and spread

» Onion bulbs and garlic cloves are most vulnerable to invasion through open sores, bruises, or unhealed neck tissue.

» Mycelium colonises the bulb and spreads through its fleshy scales before sporulating abundantly on the surface of wounds and sores.

### Management

» To ensure optimal quality, properly cure the bulbs after harvest.

» The storage area needs proper ventilation to prevent the buildup of harmful gases.

» It is recommended that bulbs be harvested with as little damage as possible, and that they be dried as soon as possible.

» The ideal conditions for storing something are a cool temperature (about 5 °C) and a high humidity level.

» The disease can be managed by treating bulbs with a fungicide.

## Soft rot, *Pectobacterium* (*Erwinia*) *carotovora* sub sp. *carotovora*

### Symptoms

» Mature bulbs are particularly susceptible to bacterial soft rot.

» At first, affected scales look wet and a bleached grey to white.

» The fleshy scales that have been invaded by the soft rot become squishy and sticky as the bulb's interior deteriorates.

» Diseased bulbs can be squeezed for a watery, putrid-smelling thick liquid near the base of the stem.

### Survival and spread

» Heavy rains and the drying of leaves are prime times for the soil-borne pathogen, as contaminated soil and agricultural wastes are the primary sources of inoculum.

» Rain, irrigation water, and insects can all spray the bacterium around.

### Management

» Insect pests, such as the onion maggot, must be managed, and overhead irrigation should be avoided if at all possible.

» Onion tops should be left to develop before harvest to reduce the risk of bulb damage.

» In order to avoid moisture condensation from accumulating on onion bulbs during storage, it is important to only store them after they have been well dried.

» Bulb storage facilities with plenty of ventilation and adequate air flow.

» Remove the spoiled light bulbs by regularly mixing up the stacks.

» Copper-based bactericides have the potential to lessen the prevalence of disease and disease.

**Iris yellow spot, *Iris Yellow Spot Virus* (IYSV) (Tospovirus: Bunyaviridae)**

## Symptoms

> » In the leaves, scape, and bulb leaves of onions and other Allium host species, necrotic lesions can range in shape from an eye to a diamond and range in colour from yellow to light green to straw.

> » Oval, concentric rings characterise lesions in the first phases of an infection.

> » Necrotic lesions contain some emerald oases.

> » In most cases, they begin near a thrips breeding ground.

> » Falling over of infected leaves occurs at the site of infection towards the end of the growing season.

> » When crops are infected when still young, it reduces their potential output.

> » Even if an infection is discovered at a later point in development, it can still degrade quality and result in substantial losses.

> » Infected fields will age quickly, turning brown, and eventually dying.

> » Viruses impair plant vitality, wreak havoc on photosynthesis, and cause bulbs to shrink in size.

## Survival and spread

> » New IYSV strains and Thrips tabaci biotypes can spread more easily when contaminated transplants of onions are moved around.

> » Due to the fact that IYSV can only be transmitted by the thrips *T. tabaci*, the spread of the virus is directly related to the distribution of both infected plants and the vector.

> » Moreover, the virus spreads and survives the winter in weeds that grow near or on protected crops.

## Management

> » Naturally occurring onion sprouts and the weeds that are harbouring them must be removed (either by tillage or herbicide).

> » Crop rotation is recommended to keep thrips populations under control because of the vectors (*Thrips tabaci*) restricted host range.

> » Use only properly inspected transplants to prevent the spread of IYSV and thrips.

» Because thrips prefer other onion colours, green-leaved cultivars are immune to IYSV.

» SAR-inducing chemicals decreased iris yellow spot by 34% in treated plants (Acibenzolar-S-methyl).

» Including UV-reflective mulch, inducing SAR with acibenzolar-S-methyl, and employing insecticides are all viable options (to preserve the natural enemies of thrips).

# References

Mishra, R. K., Jaiswal, R. K., Kumar, D., Saabale, P. R., & Singh, A. (2014). Management of major diseases and insect pests of onion and garlic: A comprehensive review. *Journal of Plant Breeding and Crop Science, 6*(11), 160-170.

Gupta, R. P., Srivastava, K. J., & Pandey, U. B. (1993). Diseases and insect pests of onion in India. In *International Symposium on Alliums for the Tropics 358* (pp. 265-270).

Amin, M., Tadele, S., & Selvaraj, T. (2014). White rot (Sclerotium cepivorum-Berk) an aggressive pest of onion and garlic in Ethiopia: An overview. *Journal of Agricultural Biotechnology and Sustainable Development, 6*(1), 6-15.

Diaz-Montano, J., Fuchs, M., Nault, B. A., Fail, J., & Shelton, A. M. (2011). Onion thrips (Thysanoptera: Thripidae): a global pest of increasing concern in onion. *Journal of economic entomology, 104*(1), 1-13.

Carr, E. A., Bonasera, J. M., Zaid, A. M., Lorbeer, J. W., & Beer, S. V. (2010). First report of bulb disease of onion caused by Pantoea ananatis in New York. *Plant disease, 94*(7), 916-916.

Shahnaz, E., Razdan, V. K., Rizvi, S. E. H., Rather, T. R., Gupta, S., & Andrabi, M. (2013). Integrated disease management of foliar blight disease of onion: A case study of application of confounded factorials. *Journal of Agricultural Science, 5*(1), 17.

Kumar, V., Neeraj, S. S., & Sagar, N. A. (2015). Post harvest management of fungal diseases in onion—a review. *International Journal of Current Microbiology and Applied Sciences, 4*(6), 737-52.

Javaid, A., & Rauf, S. (2015). Management of basal rot disease of onion with dry leaf biomass of Chenopodium album as soil amendment. *International Journal of Agriculture and Biology, 17*(1).

Steentjes, M. B., Scholten, O. E., & van Kan, J. A. (2021). Peeling the onion: Towards a better understanding of Botrytis diseases of onion. *Phytopathology®, 111*(3), 464-473.

Prowse, G. M., Galloway, T. S., & Foggo, A. (2006). Insecticidal activity of garlic juice in two dipteran pests. *Agricultural and Forest Entomology, 8*(1), 1-6.

Mishra, R. K., Jaiswal, R. K., Kumar, D., Saabale, P. R., & Singh, A. (2014). Management of major diseases and insect pests of onion and garlic: A comprehensive review. *Journal of Plant Breeding and Crop Science, 6*(11), 160-170.

Schwartz, H. F., & Mohan, S. K. (2007). Compendium of onion and garlic diseases and pests. *Compendium of onion and garlic diseases and pests.*, (Ed. 2).

Denoirjean, T., Rivière, M., Doury, G., Le Goff, G. J., & Ameline, A. (2022). Behavioral disruption of two orchard hemipteran pests by garlic essential oil. *Entomologia Experimentalis et Applicata, 170*(9), 782-791.

Monnet, Y., & Thibault, J. (2001). Diseases and pests of garlic. *PHM Revue Horticole*, (427), 50-51.

Kendler, B. S. (1987). Garlic (Allium sativum) and onion (Allium cepa): a review of their relationship to cardiovascular disease. *Preventive medicine, 16*(5), 670-685.

Srivastava, KC,* Bordia, A. &Verma, S. (1995). Garlic (Allium sativum) for disease prevention. *South African journal of science, 91*(2), 68-77.

Rahman, K., & Lowe, G. M. (2006). Garlic and cardiovascular disease: a critical review. *The Journal of nutrition, 136*(3), 736S-740S.

Rahman, K. (2001). Historical perspective on garlic and cardiovascular disease. *The journal of nutrition, 131*(3), 977S-979S.

Rana, S. V., Pal, R., Vaiphei, K., Sharma, S. K., & Ola, R. P. (2011). Garlic in health and disease. *Nutrition research reviews, 24*(1), 60-71.

# 17

# Pea

## INSECT PESTS

### Aphid, *Acyrthosiphon pisum*

#### Damage symptoms

- » Aphids, once they settle on a plant, quickly reproduce and do their usual havoc.
- » Young shoots, leaves, twigs, and pods are common targets.
- » Weaken and hinder plant growth by sucking their sap.
- » Usually this causes a decrease in pod production.
- » Aphids are well-protected from insecticide spray thanks to the inward-curling leaves.
- » Pea leaf roll virus and mosaic virus are only two examples of the diseases that aphids can transmit to plants.

#### Pest identification

- » Aphid adults have long, prominent cornicles and a soft, pear-shaped body that is green, yellow, or pink.

#### Management

- » The application of nitrogen fertilizers with discretion.
- » Pest and defence populations need to be tracked regularly in the field.
- » A mustard crop or other barrier crop could be planted around the perimeter of the land.

» Crops like maize, sorghum, and millet that grow to be many feet tall can serve as effective border plants and help keep pests at bay.

» Mist with a mixture of Bifenthrin, Dimethoate, Fenvalerate, Lambda cyhalothrin, Methomyl, and Pyrethrin.

» Carbofuran soil treatment at 1 kg active ingredient/hectare.

» Use parasitoids such as *Aphadius* sp., *Aphelinus* sp., and *Diaeretiella rapae.*

» The lacewing, ladybird beetle, predatory mite, syrphid fly, and hover fly are only few of the insects that prey on them.

» Green lacewing (*Chrysoperla zastrow sillemi*) first instar larvae should be released at a density of 10,000 individuals per hectare.

## Leaf weevil, *Sitona lineatus*

### Damage symptoms

» When plants are young and growth is sluggish, a big population of weevils may be a problem.

» The leaves of plants that have been attacked exhibit distinctive notches in a 'U' form.

» The root nodules are mostly harmed by the larvae's eating.

» In June and July, when the adults emerge, they climb plants and gnaw semicircular notches in the leaf margins.

» When weevil populations explode, they can eat the roots right out from under new plants.

### Pest identification

» The egg is 1.5 mm long and 0.6 mm wide and is a bright yellow cigar form.

» The larva develops into a 5-7 mm long legless, coiled, cream grub.

» The mature beetle is about 5 mm in length and has a brownish coloration with white, black, and grey speckles. A white abdomen with two black oval markings extends beyond the cover of the wings.

### Management

» When peas are picked early, there is less chance of the pods breaking and the peas splitting.

» Root out the unwanted plants.

- » Cruiser 5FS seed treatment (Syngenta).

- » Most of the time, phosmet spray is used to get rid of these insects.

- » Coverings made of sturdy fabric or other suitable materials can be used to shield young plants from the elements.

- » The parasitoid *Dinarmus basalis* is found in its larval stage.

## Leaf miner, *Liriomyza huidobrensis*

### Damage symptoms

- » Tunnels between the bottom and upper epidermis created by larvae prevent photosynthesis and stunt the plants' growth, making them seem unsightly.

- » Mined leaves made of serpentine.

- » In extreme circumstances, leaf drop occurs as a result of drying.

### Pest identification

- » Very little and bright orange-yellow in color; an egg.

- » Apodous maggot, as a larva. The adult maggot has a length of 3 mm.

- » The adult stage of this fly is characterized by its pale yellow colour and short length (only 1.5 mm).

### Favorable conditions

- » Increases in productivity can be seen when temperatures are high.

### Management

- » Get rid of and throw away the mined and blotched leaves.

- » To catch adult leaf miners, set up yellow sticky traps or use yellow sticky cards.

- » Parasitoids that feed on larvae and pupae include *Chrysocharis pentheus, Diglyphus isaea, Gronotoma micromorpha, Neochrysocharis formosa,* and others.

- » The green lacewing, the ladybird beetle, the spider, and the red ant are all potential predators.

# DISEASES

### Downy mildew, *Peronospora viciae*

### Symptoms

- » Plant stunting and leaf damage is the result of downy mildew.
- » The underside of the leaf develops a greyish white mould growth, while the upper side turns a sickly yellow.
- » Leaves that have been infected often turn yellow and eventually die if the weather is cool and humid.
- » There may be deformation and stunting of the stems.
- » The pods may develop brown spots, and mould may colonise their interiors.
- » Infected seedlings rarely make it if their condition is critical.
- » When the condition is severe, production is quite likely to drop.

### Favorable conditions

- » Infection and progression of the disease are facilitated by a combination of high humidity and low temperatures (5-15°C) over a few days.

### Survival and spread

- » This pathogen can be spread through seeds or the soil, and it can live there for up to 15 years.
- » Initial infection can occur by contact with soil, water, or seeds; secondary infection can occur from exposure to sporangia in the environment (whether through rain splash or wind).
- » Spore release is facilitated by heavy dews, while rain is the primary agent of spore dissemination and secondary infection.

### Management

- » The average cycle of a farm's crop rotation is three years.
- » Clearing away dead leaves and other debris after the growing season the use of virus-free seed stock.
- » Any infected vegetation found in the field should be promptly destroyed by fire.
- » Some of the protection against soil and seed-borne diseases can be provided by Metalaxyl (Apron FL).

**Powdery mildew,** *Erysiphe polygonum*

## Crop losses

» It does a lot of harm and can reduce pod quality and quantity by 20 %.

## Symptoms

» Mycelium, a powdery greyish white growth, affects leaves initially and causes tiny spots of discoloration.

» A white powdery growth covered the plant's leaves, stem, and pods.

» The foliage begins to wither and turn yellow.

» Neither do the fruits set, nor do they grow past a tiny size.

» The result is a defoliated landscape.

» At later stages, the stems, branches, pods, and tendrils are all coated in a powdery growth.

» As a result of an early attack, plants may not produce fruit or may produce pods that are bitter.

## Favorable conditions

» Late in the growing season, when flowers are opening and pods are filling, temperatures of 15 to 25 °C and humidity levels of over 70 % are ideal for the spread of disease.

## Survival and spread

» The pathogen can live off of pea stubble and other agricultural wastes.

» As powdery mildew spores become active, they can travel through the air and continue to spread even when humidity levels are low.

## Management

» Where possible, contaminated pea stubble should be burned immediately after harvest.

» Don't plant your peas next to the remains of last year's crop.

» Eliminate any stray field peas from your garden, as they may be disease carriers.

» Use either 200 grammes of Benomyl 50% WP or 250 grammes of Carbendazim 50% WP diluted in 600 litres of water per hectare for spraying. Repeat spraying in 15 days.

» The seedling stand was significantly improved by biopriming pea seeds with *Trichoderma harzianum* and *Pseudomonas fluorescens.*

» The prevalence of powdery mildew disease was decreased and pea yield was raised after bioagents were applied via foliar spray.

» All of the Rachna, Pant P5, Ambika, Shubhra, Aparna, Azad P4, and Pusa Panna varieties are resistant to powdery mildew.

## Rust, *Uromyces pisi*

## Symptoms

» Infected plant leaves typically have numerous tiny, orange-brown pustules on their undersides.

» Damaged leaves wither and fall off the plant.

» Some pods may also have larger pustules, and the stems may have a few smaller ones.

» Severe infection can diminish seed size, which in turn can lower output by as much as 30 %.

## Favorable conditions

» Rainfall, many dews, and a climate of 20-25 °C all contribute to the spread of disease.

» Conditions of drought and heat slow the spread of the disease.

## Survival and spread

» It spreads through euphorbia and decaying plant matter that is contaminated.

» This disease cannot be spread through planting seeds.

## Management

» After harvesting, remove and dispose of any plant material that could harbour disease.

» Make sure to rotate in some non-leguminous crops every so often.

» Variegated harvesting.

» Use 3 kg of 80% WP sulphur or 0.1% of 25% WP triadimefon in 750 litres of water per hectare. After another 25 days, you can spray again.

» Use 2 kg of Dithane M-45 per hectare and spray the crop with 1,000 litres of the solution. You just need two or three sprays spaced out by 10 days.

» Unlike other Hans, the DMR 11 Type 163 is not susceptible to rust.

## *Ascochyta* leaf and pod spot, *Ascochyta pisi*

### Symptoms

» Under the canopy, on lower leaves, stems, and tendrils, where circumstances are more humid, early signs are most often seen.

» Little, uneven, purplish-brown specks are the initial sign of the disease.

» In the presence of persistent moisture, the specks grow and combine, eventually causing total blight of the lowest leaves.

» Girdling of the stem just above the soil line, often known as foot rot, can be caused by a bacterial infection of sufficient severity.

» Lesions caused by foot rot are often a dark purple or black in colour and can spread both on the surface of the ground and deeper into the soil.

» Crop lodging and yield loss can be caused by lesions on the foot and stem, which girdle and weaken the stem.

» Disease lesions appear on pods when the crop is wet for an extended period of time or when it becomes stuck.

» Little, black spots on the pods are the first sign that something is wrong, but these lesions can spread and cause premature pod senescence if not treated.

» Seeds from infected pods may be abnormally tiny, shrunken, or discoloured, or they may show no symptoms at all.

### Favourable conditions

» Climate conditions are warm and humid, with temperatures ranging from around 15o C to 25°C.

### Survival and spread

» May overwinter in plant remains or enter new environments in diseased pea seeds.

» Spores are carried by wind and rain to healthy plants.

» The asexual conidia are carried by raindrops to nearby hosts.

## Management

» As soon as signs of disease show, the affected plants should be removed and destroyed.

» Switch up your sensitive crops every year for some that aren't.

» The sowing of seeds that have been tested and found to be disease-free.

» Crops of tall plants are planted to prevent the spread of infection through the air.

## White rot, *Sclerotinia sclerotiorum*

## Symptoms

» There is no one section of the leaf that is more susceptible to infection than any other.

» The peak of the infection occurs when the crop is in full bloom and its petals fall to the ground, where they are quickly colonised by the fungus and from which mycelial growth invades the stem and branches.

» A dry, brownish patch appears at the site of infection.

» It starts low and works its way up and down the stem.

» Tissue necrosis causes the healthy plant tissue beyond the infected area to die off.

» If the plant's root system is infected, the entire plant will die.

» The affected branches may begin to partially wither.

» Infected tissues turn white and may tear to pieces.

## Favorable conditions

» Sclerotitis is more likely to occur in humid environments with dense canopies.

» When temperatures are between 4 and 20 °C and there is sufficient moisture in the air.

» Apothecial production must have light in order to be stimulated.

## Survival and spread

» Sclerotial infection is an disease brought on by infected sclerotia that have been left in the soil or on dead plants.

## Management

» Maintain larger gaps.

» Regulate watering.

» Four or five years without growing any vulnerable crops is ideal (canola, mustard, sunflower, dry bean).

» Benlate 50WP (Benomyl) and Rovral Spray (Iprodione).

## Root rot, *Aphanomyces euteiches* f. sp. *pisi*

## Symptoms

» The lower stem and roots gradually generate long, pliable, water-soaked patches that eventually turn tan and extend throughout the root system.

» Roots of highly diseased peas are simple to remove from the ground and peel away, revealing the inner tissues beneath.

» Diagnostic oospores (25-30 microns with an oil globule in the middle) are easily observed in cortical tissues when using a compound microscope.

» Infected plants' upper parts become stunted and, in extreme cases, wilt, yellow, and die before their time.

## Favorable conditions

» Drizzle and cloud cover, meaning it's cool outside.

» Increased heat and moisture in the soil, between 25 and 30 degrees C.

## Survival and spread

» Transmission of the thick-walled oospores that overwinter the pathogen can occur by water, wind, infected plant debris, soil movement, or tillage machinery.

» The primary vectors of transmission are water, wind, infected plant debris, soil movement, and tillage machinery.

» Secondary dispersal of conidia via wind or splashes of rain.

## Management

» The disease is best controlled by lengthy rotations.

» By reducing inoculum levels, green manure crops (oat and Brassica residues) can be grown and incorporated into soil.

» Choose land that doesn't flood easily.

» Stay away from areas where the dirt is too compacted.

» You should stay away from the infected areas.

» Find out how much inoculum is present in a field by doing soil tests.

» Planting late-maturing types in soils that have had root rot in the past is not recommended.

## Fusarium wilt, *Fusarium oxysporum* f. sp. *pisi*

### Symptoms

» Most noticeable disease symptoms appear in plants between the ages of 3 and 5 weeks.

» Infected plants show symptoms like drooping and withering during a vulnerable time in their development.

» Lower leaves turning yellow and plant growth stunted.

» The stem's vascular darkening and pith's reddish color appear to have travelled all the way down to the roots.

» If the roots aren't removed, they'll turn black and decay.

» The plant's development slows, its leaves turn yellow, and the stipules and leaflets curl downward.

» When the plant's stem shrivels, the whole thing dies.

» A common symptom is the development of epinasty in afflicted plants.

### Favorable conditions

» The ideal soil temperature for Fusarium wilt is between 23 and 27 °C.

» High air and ground temperatures are ideal for plant growth.

### Survival and spread

» Infection at the primary stage, spread by the environment.

» Conidia spread through rain splash to a secondary host.

### Management

» Solarization as a means of soil sterilisation.

» In heavily infested locations, early sowing should be avoided.

» Thiram (3 g/kg of seed) or a 0.1% Bavistin solution (for 30 minutes) applied to seeds.

» Roots of infected plants are cut off, which lessens the spread of the disease.

» Crops should be rotated regularly.

» Incorporation of green manure into soil.

» The use of *Trichoderma harzianum* is quite advantageous.

## Bacterial blight, *Pseudomonas syringae* pv. *pisi*

### Symptoms

» Early symptoms of bacterial blight on leaves and stipules are water-soaked, dark green patches that tend to cluster near the leaf base. Spots swell and join, but their expansion is typically halted by nearby veins.

» Yellow dots on the leaves become dark and papery as time passes.

» The spots on the pods are recessed and olive green in colour.

» Patches can appear on the lower part of the stem. These regions start out looking waterlogged and eventually transform from olive green to dark brown. The stem could shrivel and perish if the lesions accumulated.

» The stipules and leaflets are at risk if lesions extend up the stem. On the stipule, a fan-shaped lesion has developed.

» Pods are not immune to this disease, either.

### Favourable conditions

» Bacterial blight is at its worst during rainy seasons since the disease thrives in damp environments.

» The disease spreads most rapidly within crops when there is both heavy precipitation and high winds.

### Spread

» Rain splash and wind-borne water droplets allow bacteria to travel from sick to healthy plants during wet weather.

### Management

» Transplanting healthy seedlings.

» Turning over your crops every so often. Growing peas on the same plot of land more than once every three years is not recommended.

» Inbreeding for drought resistance is a viable option.

» Not planting seeds too soon.

» Maintain clean conditions, ensure enough drainage, and space your plants appropriately.

» Kill the sick weeds by burying, baling, or burning them.

» Streptomycin (0.01%) spray should be applied at the first sign of disease and thereafter, if necessary, every 7 days.

## References

Singh, S. R., & Emden, H. V. (1979). Insect pests of grain legumes. *Annual review of Entomology, 24*(1), 255-278.

Hussain, R., Ihsan, A., Shah, A. A., Ullah, N., Iftikhar, H., & Jalal, R. (2022). Ecofriendly Management of Green Pea (Pisum sativum L.) Insect Pests through Plant Extracts. *Sarhad Journal of Agriculture, 38*(4), 1405-1411.

Biddle, A. J., & Cattlin, N. (2007). *Pests, diseases and disorders of peas and beans: A colour handbook.* CRC Press.

Clement, S. L., El-Din Sharaf El-Din, N., Weigand, S., & Lateef, S. S. (1993). Research achievements in plant resistance to insect pests of cool season food legumes. *Euphytica, 73,* 41-50.

Tomar, S. P. S., Dubey, O. P., & Rajani, T. (2004). Succession of insects pest on green pea. *JNKVV Research Journal, 38*(1), 82-85.

Ransom, L. M., O'Briens, R. G., & Glass, R. J. (1991). Chemical control of powdery mildew in green peas. *Australasian Plant Pathology, 20*(1), 16-20.

Basu, P. K., Crete, R., Donaldson, A. G., Gourley, C. O., Haas, J. H., Harper, F. R., ... & Zimmer, R. C. (1973). Prevalence and severity of diseases of processing peas in Canada, 1970-71. *Can. Plant Dis. Surv, 53*(4).

Hidayat, M., Prahastuti, S., Yusuf, M., & Hasan, K. (2021). Nutrition profile and potency of RGD motif in protein hydrolysate of green peas as an antifibrosis in chronic kidney disease. *Iranian Journal of Basic Medical Sciences, 24*(6), 734.

Fondevilla, S., & Rubiales, D. (2012). Powdery mildew control in pea. A review. *Agronomy for sustainable development, 32,* 401-409.

SM, M. M., & Golshani, B. (2013). Simultaneous determination of levodopa and carbidopa from fava bean, green peas and green beans by high performance liquid gas chromatography. *Journal of clinical and diagnostic research: JCDR, 7*(6), 1004.

# 18

# Pigeon Pea

## INSECT PESTS

### Pod borer, *Helicoverpa armigera*

#### Damage symptoms

» For a brief period of time, larvae consume sensitive plant parts like leaflets, flower buds, and young roots.

» It slowly burrows into pods, where it feasts on the seeds.

» When feeding on growing seeds, half of the larva stays within.

» Clearing vegetation during an infestation's first stages.

» Round-bored pods.

#### Pest identification

» The eggs are round, white, and laid individually.

» The larvae can range in color from a bluish-green to a tan brown. The body is green, and there are dark brown grey lines running laterally, as well as white lines, and dark and pale bands.

» Brown pupae can be found in dirt, leaves, pods, and other plant matter.

» Adults of this species are a pale brownish yellow in colour and quite stocky in build. Wingtip specks are V-shaped, and range in colour from grey to pale brown on the front. The outermost margin of the wings behind the back are a dark bluish black.

## Management

» Plant 9 rows of pigeon peas and 1 row of sunflowers as intercrops, with corn planted as a border crop.

» There will be 12 pheromone traps per hectare to catch *Helicoverpa armigera*.

» A total of 50 bird feeders per hectare.

» Pigeon pea is less susceptible to *H. armigera* when it is intercropped with sorghum and with short-season legumes like soybean or green gramme.

» Shaking pigeon pea plants by hand to expel larvae is a common practise in India during times of significant infestation.

» Neem oil (2%), Azadirachtin (0.03%) WSP (2.5-5.0 kg/ha), or both.

» Neem cake (250 kg/ha) worked into the soil just before flowering increased yields by 11%.

» Use 5% NSKE twice, then 0.05% Triazophos.

» Emamectin benzoate (5% SG 220 g/ha), Indoxacarb (15.8% SC 333 ml/ha), Chlorantraniliprole (18.5 SC 150 ml/ha), Spinosad (45% SC 125-162 ml/ha), and Phosalone (0.07%) are all pesticides that can be sprayed.

» Spray HaNPV at 3 x 1012 POB/ha in 0.1% Teepol or *Nomurea rileyi* with *Bacillus thuringiensis* serovar kurstaki (3a,3b,3c) 5% WP.

» Nematode DD-136 (*Steinernema feltiae*) at 3 103 infective juveniles/ml and HaNPV at 2 106 PIBs/ml for integrated management.

## Plume moth, *Exelastis atomosa*

## Damage symptoms

» The pods' openings are the size of pinheads.

» There are tiny spiny caterpillars and their pupae on the pods.

» The buds, flowers, and pods are damaged by larvae feeding on them.

» The buds and young pods have tiny holes in them.

## Pest identification

» Oval and green, eggs are a symbol of fertility.

» The larva has a fringed, spindle-shaped body and is covered in small hairs and spines that give it a greenish brown colour.

» The adult moth is a tiny brown butterfly with delicate, feathered wings.

## Management

- » Crop planting at the right time.
- » Incorporation of non-host crops into a farmer's rotation.
- » Keeping the playing field spotless.
- » Planting crops that aren't hosts for pests.
- » We need to get rid of and dispose of the damaged plant parts.
- » Insecticide spraying with Monocrotophos 36 SL at 2 ml/L of water at the 50% flowering stage is recommended to prevent damage from moths and larvae.

## Spotted pod borer, *Maruca testulalis*

## Damage symptoms

- » Cut holes in the developing flower, pod, or bud.
- » Pods and blossoms infected with the pest web together.

## Pest identification

- » The larva is a whitish green colour and has a brown head. Its back segments are marked by two sets of dark dots.
- » The adult moth has white patterns on its light brown forewings and brown markings along its white hind wings.

## Management

- » The early part of July is the absolute latest you should sow.
- » Application rates of nitrogen should be monitored closely and excess amounts avoided.
- » Avoiding flooding is of paramount importance.
- » Crop rotation involving corn, cowpeas, and sorghum.
- » The effectiveness of biocides like *Bacillus thuringiensis* (Bt) and Neem products like Neem seed kernel extract (NSKE) or Neem oil varied.
- » The Spinosad 45% SC formulation at 73 g a.i./ha had the least amount of inflorescence and pod damage from the legume pod borer, followed by the *Bacillus thuringiensis*-1 at 1.5 kg/ha and the *Beauveria bassiana* SC formulation at 300 mg/L.

» Use 600-1000 L of spray material per hectare using 0.07% Endosulfan (2 ml of 35 EC/L of water), 0.04% Monocrotophos (1 ml of 36 SL /L of water), or Chlorpyriphos 20 EC (3.5 ml/L of water).

» Protect species like ants and praying mantids, which prey on pest eggs and larvae.

» Antrocephalus sp., an endoparasitoid that attacks pupae, was the most common predator, whereas *Nosema* sp. and *Bacillus* sp. were responsible for the highest rates of natural mortality.

## Pod fly, *Melanagromyza obtusa*

### Damage symptoms

» Walls of the pod were covered in a dark brown encrustation.

» Pods were dry and had a hole the size of a pinhead.

» The larvae eat the green seed from the inside out.

» Slimy, spotted, and half devoured seeds.

### Pest identification

» Little, black fly that matures into a pest.

» The 3-mm-long, whitish larvae have no legs.

» Color of puma fur.

### Management

» The practise of planting different crops alongside one another, such as sorghum and maize or groundnuts.

» Planting crops that aren't hosts for pests.

» Systemic/contact insecticides with fumigant action (Monocrotophos at 1.5 ml/L/ Nuvan at 1.0 ml/L) can be used to effectively reduce leaf webbers in cases of severe infestation.

» This pest is also linked to a wide variety of parasites and predators (*Coccinella septumpunctata*, *Mentispa indica*, Apanteles, Spiders and Rove beetles).

» The Neem seed kernel extract should be applied 10–15 days after the 50% blooming stage was treated with either Monocrotophos @ 36 SL @ 1 ml/L or Endosulfan 35 EC 2.0 ml/L of water.

» Acephate 75SP 1g/L water for spraying.

» Dimethoate and Monocrotophos are systemic insecticides that can be used to control the larvae that feed on the seeds and pods from the inside.

» Endosulfan, a non-systemic pesticide, is useful for eliminating adult mosquitoes.

» Use a spray containing either 25 milligrammes of thiamethoxam, 17.8 milligrammes of imidacloprid, or 20 milligrammes of dinotefuran.

» Keep the pod fly parasite *Ormyrus* sp.

## Leaf webber, *Grapholita critica*

### Damage symptoms

» As they develop, larvae use silk to sew together leaflets.

» In the centre of a web of leaves, flowers, and pods is where they eat.

» Infestation of the terminal bud can significantly limit the development of a whole shoot.

» It is possible for infestations to start when the seeds are still in the podging stage and last all the way through the blooming and fruiting phases.

» *G. critica* is mostly a leaf eater, but it can also damage reproductive structures if it gets in late enough.

» The leaf webber acts like a pod borer when conditions are right.

### Pest identification

» A mature larva's anterior (thoracic segments) was lighter in colour than its posterior (3-4 abdominal segments).

» The pupae were initially a pale yellow colour, but they darkened to a reddish brown during time.

» The adult stage of this species is characterised by a brown moth with a wing span of only 10-15 mm.

### Management

» Having cereals and legumes in the same field together greatly impacted the decline of the leaf webber larval population.

» In order to control leaf webbers, it was observed that two sprayings with a mixture of NSKE (5%), Nimbecidine (1%), and *Bacillus thurungiensis* var. *kurstaki* (1.5%) were highly efficient.

» The most successful treatments were found to be spraying with Monocrotophos 0.05%, Quinalphos 0.05%, Fenvalerate 0.015%, Endosulfan 0.07%, and Profenophos 40 EC + Cypermethrin 4 EC (0.044%).

» The damage caused by leaf webber was less severe on the BSMR-736 cultivar.

# DISEASES

**Fusarium wilt,** *Fusarium udum*

**Symptoms**

» When the crop is flowering or podding, the first sign of wilt is a cluster of dead plants in the field.

» A purple ring rising from the lower part of the main stem is the most telling indication.

» Chlorosis turns leaves a vivid yellow and causes a loss of turgidity and interveinal clearance.

» *Fusarium wilt* can be identified by the telltale symptom of partial plant withering.

» Wilting at the edges is caused by an infection in the plant's lateral roots, whereas wilting at the centre is caused by an infection in the tap roots.

» When the main stem or primary branches are cut apart, you can see that the tissue around the purple band has darkened in colour, and the xylem has darkened to a brown or black color.

**Favorable conditions**

» Comparatively, Vertisols have a higher disease incidence than Alfisols.

**Survival**

» Disease spreads through soil and seeds.

» This fungus has a three-year life span when left on infected plant detritus in the soil.

**Management**

» The wilt resistance of pigeon peas is increased when they are intercropped with sorghum.

» Dressing seeds with 3 g of Benlate T (Benlate 5% + Thiram 50%).

» Clear polythene sheet of 400 gauges (94 g/ m² and 100 m thick) is used for solarizing soil.

» ICP 8863 (Maruti), Birsa Arhar 1, Mukta, Prabhat, and Sharda are wilt-resistant cultivars to grow.

## Collar rot, *Sclerotium rolfsii*

### Symptoms

» There are areas of the field where young, primary-leaf-stage seedlings have died.

» Before they finally perish, the seedlings may get a mild chlorotic color.

» A rotting collar area covered in white mycelial growth is diagnostic.

» Seedlings with collar rot are easily uprooted, although their lower roots typically remain in the ground.

» Fungal sclerotial bodies, either white or brown, may be observed clinging to the collar region of a deceased seedling, or present in the soil nearby.

### Favorable conditions

» As a general rule, germinating seeds are most vulnerable to disease when temperatures are about 30°C and soil moisture is high.

### Management

» Choose areas that didn't host cereal crops (like sorghum) the year before.

» To prepare a field for planting pigeon peas, you must first collect and eliminate any grain stubbles that may be present.

» Tolclofos-methyl (3 g kg$^{-1}$) seed dressing (Rhizolex), Captan (3 g kg-1 seed), or Thiram (3 g kg$^{-1}$ seed) are all viable options.

## Powdery mildew, *Levillula (Oidiopsis) taurica*

### Symptoms

» Each and every aerial component of infected plants, including the leaves, flowers, and pods, is covered in a white, powdery fungal growth.

» Heavy defoliation occurs when infections are severe.

» Young plants are stunted by the disease, and then, just before they bloom, a white powdery growth appears, a visible indication of the sickness.

» Little chlorotic spots appear first on the upper surface of affected leaves, followed by white powdery patches on the undersides.

» The fungus produces a white powdery growth that coats the underside of the entire leaf when it sporulates.

## Management

» Avoid planting near infected perennial pigeon pea in your chosen fields.

» Planting after July in India is recommended in order to delay the onset of disease.

» Inject 0.03% Triadimefon (Bayletan 25% EC) or 1 g L⁻¹ Wettable Sulfur.

» Experiment with the ICP 9150 and ICP 9177 resistant variants.

## Dry root rot, *Rhizoctonia bataticola, Macrophomina phaseolina*

## Symptoms

» Infected plants wither and die quickly.

» The roots of such plants are rotting and shred when uprooted.

» Roots with thinner bark are most severely impacted; the dark sclerotial bodies beneath the bark are easy to see.

» Roots like these are fragile and easily damaged. When temperatures and humidity levels are high, root rot can spread all the way to the stem's base.

» Spindle-shaped lesions with light grey cores and brown edges and scattered pycnidial bodies are the earliest signs to appear on stems and branches.

» The lesions join together, eventually killing off entire branches or even plants.

## Favorable conditions

» Infected plants wither and die quickly.

» The roots of such plants are rotting and shred when uprooted.

» Roots with thinner bark are most severely impacted; the dark sclerotial bodies beneath the bark are easy to see.

» Roots like these are fragile and easily damaged. When temperatures and humidity levels are high, root rot can spread all the way to the stem's base.

» Spindle-shaped lesions with light grey cores and brown edges and scattered pycnidial bodies are the earliest signs to appear on stems and branches.

» The lesions join together, eventually killing off entire branches or even plants.

## Management

» Go for areas where dry root rot has never been a problem before.

» Late planting should be avoided if the crop is to mature without being damaged by drought or excessive heat.

» Use a *Trichoderma viride* formulation that had 4 g per kg of seed and 3 g of Thiram, and we applied 2 kg of *T. viride* formulation blended with 125 kg of FYM per hectare.

## *Phytophthora* blight, *Phytophthora drechsleri* f. sp. *cajani*

### Symptoms

» Phytophthora blight is similar to damping off disease in that both of them result in the abrupt death of seedlings.

» Diseased plants' leaves will have water-soaked lesions, while the stems and petioles will develop brown to black, somewhat sunken lesions.

» Infected leaves dry out and lose their pliancy.

» The lesions encircle the damaged main stems or branches, causing them to break and the foliage above the lesion to die.

» When environmental factors favour the disease, widespread plant mortality often results.

» Large galls form on the stems of infected, but surviving, pigeonpea plants, especially at the lesional margins.

» The root system is resistant to the disease, but the plant's foliage and stems are susceptible to infection.

### Favorable conditions

» Infections that necessitate a constant dampness of the leaves for 8 hours are more likely to occur under conditions of cloudy weather and drizzling rain with temperatures around 25°C.

### Survival and spread

» This disease spreads through the soil. The fungus can persist in soil and on infected plant detritus as chlamydospores, oospores, and quiescent mycelium.

» After infection, disease progresses rapidly in warm and humid conditions, killing off plants. In addition to the help of wind and rain, zoospores can also travel great distances.

## Management

- » Mefenoxam seed treatment alone, Mefenoxam seed treatment with Trichoderma isolate-3, Mefenoxam seed treatment with Pseudomonas isolate-1, and Metalaxyl seed treatment alone were all successful.
- » Supplementing with *Trichoderma* and *Pseudomonas* can lower the number of fungicides needed for disease management without sacrificing their effectiveness.

## Sterility mosaic disease, *Sterility Mosaic Virus*

## Symptoms

- » Affected areas are characterised by dense clusters of bushy, light-green plants that lack both flowers and pods.
- » This plant has tiny leaves with a mosaic pattern of lighter and darker green.
- » First, you can notice vein clearing on young leaves, an indication of mosaic disease.
- » Infection at 45 days post-emergence or later may cause localised symptoms while leaving other portions of the plant unaffected.

## Favorable conditions

- » Mites flourish in cool, damp environments, which is why they thrive in the shade during the warmer months.

## Survival and spread

- » The mite vector and the pathogen can be stored in reservoirs such as perennial and volunteer pigeon peas and the ratooned growth of harvested plants.
- » One eriophyid mite is all it takes to spread the disease. They can be carried by the wind up to 2 kms from their point of origin.

## Management

- » At the first sign of disease, remove and kill any affected plants.
- » Crop rotation can lessen the likelihood of mites and other pests spreading from plant to plant.
- » Use 3 grammes per kg of seed of either 10% phospate or 25% furadan 3G as a seed treatment.
- » Controlling the mite vector in the early phases of plant growth with the application of acaricides or insecticides such as Kelthane, Morestan, and

Metasystox at 0.1% through spraying.

» Bageshwari, Bahar, and Rampur Rahar, which are disease-resistant.

» The ICPL 151 cultivar, known as Jagriti, is resistant to both the Sterility mosaic virus and the Root-knot nematode.

# References

Shanower, T. G., Romeis, J. M. E. M., & Minja, E. M. (1999). Insect pests of pigeonpea and their management. *Annual review of entomology*, *44*(1), 77-96.

Hillocks, R. J., Minja, E., Mwaga, A., Nahdy, M. S., & Subrahmanyam, P. (2000). Diseases and pests of pigeonpea in eastern Africa: a review. *International Journal of Pest Management*, *46*(1), 7-18.

Sharma, O. P., Gopali, J. B., Yelshetty, S., Bambawale, O. M., Garg, D. K., & Bhosle, B. B. (2010). Pests of pigeonpea and their management. *NCIPM, LBS Building, IARI Campus, New Delhi-110012, India*, *4*(1080), 07352681003617483.

Dialoke, S. A., Agu, C. M., Ojiako, F. O., Onweremadu, E., Onyishi, G. O., Ozor, N., ... & Ugwoke, F. O. (2010). Survey of insect pests on pigeonpea in Nigeria. *Journal of SAT Agricultural Research*, *8*, 1-8.

Lateef, S. S., & Reed, W. (1983). Review of crop losses caused by insect pests in pigeonpea internationally and in India. *Indian Journal of Entomology*, *2*, 284-293.

Singh, R., Singh, B. K., Upadhyay, R. S., Rai, B., & Lee, Y. S. (2002). Biological control of Fusarium wilt disease of pigeonpea. *The Plant Pathology Journal*, *18*(5), 279-283.

Singh, V. K., Khan, A. W., Saxena, R. K., Kumar, V., Kale, S. M., Sinha, P., ... & Varshney, R. K. (2016). Next-generation sequencing for identification of candidate genes for Fusarium wilt and sterility mosaic disease in pigeonpea (C ajanus cajan). *Plant Biotechnology Journal*, *14*(5), 1183-1194.

Pande, S., Sharma, M., & Guvvala, G. (2013). An updated review of biology, pathogenicity, epidemiology and management of wilt disease of pigeonpea (Cajanus cajan (L.) Millsp.). *Journal of Food Legumes*, *26*(1and2), 1-14.

Butler, E. J. (1910). The wilt disease of pigeonpea and the parasitism of Neocosmospora vasinfecta Smith. *The wilt disease of pigeonpea and the parasitism of Neocosmospora vasinfecta Smith.*, *2*, 1-64.

Gnanesh, B. N., Bohra, A., Sharma, M., Byregowda, M., Pande, S., Wesley, V., ... & Varshney, R. K. (2011). Genetic mapping and quantitative trait locus analysis of resistance to sterility mosaic disease in pigeonpea [Cajanus cajan (L.) Millsp.]. *Field Crops Research*, *123*(2), 53-61.

# 19

# Potato

## INSECT PESTS

### Tuber moth, *Phthorimaea operculella*

#### Damage symptoms

- » The larva of the potato tuber moth bores into tubers, stems, and leaves.
- » Storage areas are common places for a severe infestation to manifest, leading to the formation of crooked galleries and "tunnels" well into the tube.
- » In the area of bore holes, you can find the excrement of larvae.
- » They burrow into the leaves, devour the pulp, and exit, leaving behind only the shrivelled shell.

#### Pest identification

- » Moth with dirty white hind wings and greyish brown forewings as an adult.
- » The caterpillar stage is represented by a yellow body and a dark brown head.
- » The eggs are laid singly on the underside of the leaves and exposed tubers.

#### Management

- » Deep planting, high ridging, and cool storage of seed are all effective pest management strategies.
- » Choose robust tubers for planting.
- » Don't put tubers in a hole that's too shallow. Put the tubers in the ground about 10 to 15 cms deep.
- » Placing 15 pheromone traps per hectare is recommended.

» The diseased tubers in the field must be gathered and destroyed.

» Don't go to bed with the tubers still out in the field.

» Try growing peppers, onions, and peas together as an intercrop.

» To prevent female moths from depositing eggs on your exposed tubers, dirt up about day 60 after planting.

» To prevent adult moths from laying eggs on the eyes of the tubers, keep them in storage before sunset covered with a 5 cm sand layer.

» To keep the potato tuber storage area free from ovipositing moths, lay down a layer of Lantana and Eupatorium branches on top of the tubers.

» Sex pheromones on water traps at a ratio of 4 traps per 100 m3 of storage capacity are effective for mass catching of males.

» *Chelonus blackburni, Copidosoma koehleri, Trichogramma* spp., *Apanteles sp., Pristomerus vulnerator,* and many others are all examples of parasitoids.

» Insects like the lacewing, red ant, ladybird beetle, spider, robin fly, dragonfly, and others are predators.

» Usage of biological pesticides like *Bacillus thuringiensis* and the Granulose Virus.

» The parasitoid Chelonus blackburnii should have its eggs and larvae released at a rate of 30,000 per hectare at 40- and 70-days post-planting.

» Manage foliar damage by spraying NSKE at 5% or Quinalphos 25 EC at 2 ml/L in water.

## Aphid, *Myzus persicae*

### Damage symptoms

» Aphids are a pest because they puncture plants and drink their sap.

» They are particularly destructive to the tender, immature parts of plants, like the leaves and shoots.

» Not opening fully and shrinking are both symptoms of harm to leaves.

» Weakened and dried-out shoots are a common result of a severe infestation.

» Aphids are winged insects that can carry plant viruses from one plant to another.

» Black sooty mould thrives in environments where aphids have secreted a sugary liquid.

» It can coat the plants and reduce their photosynthesis by blocking the sun.

**Pest identification**

» The abdomen of winged (alate) aphids is yellowish green with a huge dark patch on the dorsal side.

» There are many similarities between a nymph and an adult, however the nymph is much smaller.

**Favorable conditions**

» These aphids thrive in temperatures between 11 to 14°C.

» Over a range of relative humidity (RH) of 73% and above, its population drops precipitously.

**Management**

» Planting at times when aphid populations are low or not present at all can increase yields.

» You should get rid of any weeds that could potentially harbour viruses or aphids.

» As soon as the aphid population reaches the threshold level, defined as 20 aphids per 100 leaves, the haulm cutting of the seed crop must be performed.

» Use yellow sticky traps to keep an eye on aphid swarms.

» You should use a 0.3% solution of dimethoate in spray form.

» The release of predators and parasitoids (such as ladybird beetles, lacewings, spiders, hover flies, and the like) is recommended (*Lysiphlebus* sp, *Diaeretiella* sp., *Aphelinus* sp., *Aphidius colemani*, etc.).

**Whitefly, *Bemisia tabaci***

**Damage symptoms**

» Nymphs sap the vitality of plants by feeding on their juices.

» Leaf withering and wilting.

» Leaves are drying and curling downward.

» Honeydew secreted by whiteflies feeds sooty fungi, further damaging the leaves.

» As a result, such factors have a negative impact on photosynthesis in plants, which in turn reduces crop yields.

» Insects that spread potato leaf curl.

## Pest identification

» Eggs are round or slightly elliptical, with a broad base and a stalk. When first laid, it's a pale yellow, but it eventually turns a dark brown.

» The stalked nymph is a slow-moving, pale-yellow organism with the appearance of a louse.

» Pupae are rounded and have dark yellow markings on their bellies.

» Fly adults are tiny flying insects measuring just 1.0-1.5 mm in length and covered in a fine, white waxy substance all over their bodies. White and elongated, the wings' legs stand out against the background.

## Favorable conditions

» This pest thrives in the dry heat of August to January, when temperatures range from 28 to 36°C and humidity levels from 62 to 92%.

## Management

» Excellent watering timing.

» In endemic regions, you should stay away from common Solanaceous crops.

» Imidacloprid seed treatments and subsequent foliar applications, the latter of which is done 15 days following emergence, are recommended.

» A few examples of parasitoids are the nematode Encarsia formosa and the cestode Eretmocerus species.

» Ladybird beetles, lacewings, spiders, hover flies, reduviid bugs, robber flies, and many more insects and animals serve as predators.

» Kufri Bahar potatoes are very resistant to whitefly.

## Leaf hopper, *Empoasca kerri*

## Damage symptoms

» Adult and juvenile jassids feed on plant sap from the undersides of leaves.

» The sucking action causes the plants to become stunted, the leaf edges to curl, and the upper surfaces of the leaves to become wrinkled.

» During eating, leafhoppers inject plants with a poisonous chemical, resulting in "hopper- burn" in most vegetable hosts.

» The disease causes the tissue at the leaf tips and leaf margins to yellow and eventually die.

## Pest identification

» Eggs are of an elongated, yellow-white colour. Before they hatch, they turn a drab grey-yellow.

» The wing pads of nymphs, which are a whitish green in colour, extend to the fifth abdominal segment.

» The full-grown specimen is a pale green bug with a black mark on each of its see-through forewings. In the winter, it takes on a rusty brown color. The average lifespan of a fully grown adult with wings is 35-50 days.

## Favorable conditions

» The ideal conditions for jassid reproduction include a temperature range of 27 to 36°C and a relative humidity of below 75%.

## Management

» Imidacloprid can be used to treat seeds.

» In order to effectively control early leaf hopper larvae and adults, neem seed extract (1 kg seed powder/40 litres of water) was sprayed three times at 10-day intervals.

» Phorate is sprayed on the soil before planting.

» At the first sign of an infestation, spray a solution of 2 ml of phosphomidon (Dimecron) or dimethoate (30 EC) or methyl demeton (Metasystox 55) per litre of water.

» Application of Anagrus flaveolus, Stethynium triclavatum, and other parasitoids.

» The lacewing, red ant, mirid bug, big-eyed bug, ladybird beetle, and many more insects are predators.

» The potato leafhopper doesn't have a chance against the 'Delus' strain.

## Cut worm, *Agrotis ipsilon, A. flammatra*

## Damage symptoms

» During night, these pests do their damage to plants and tubers.

» Larvae eat the outer layer of leaves.

» They slash the stems of tender young plants, drag them into the soil, and eat everything save the roots.

» There is little hope for a plant's recovery if its stem has been damaged.

» Even at 25–35 days post-planting, this bug can do significant harm to crops.

» Bore holes in tubers that are bigger than those formed by potato tuber moths are an indication of damage.

## Pest identification

» Eggs: Round, white, and creamy, placed singly on the undersides of leaves.

» Larvae: The immature larvae that have just emerged are a bright yellow. The mature larva has a swollen, oily body and is a dark brown colour.

» Dark brown pupae can be observed in underground cells in potato farms.

» The adult moth is mostly black, with a few greyish patches on the back and a few black streaks on the forewings.

## Favorable conditions

» Cutworms thrive in conditions like low humidity, temperatures between 16 and 23°C, and extended periods of dry weather.

## Management

» Kill the pests by drowning them in flooded fields.

» Pick the larvae out of the cracks and crevices in the field in the early morning and late evening and eliminate them.

» In the summer, ploughing the ground will expose the larvae and pupae to the birds that will eat them.

» To catch male moths, set out pheromone traps at a rate of 12 per hectare.

» Pesticides such as Chlorpyriphos 20 EC at 1 litre per hectare (L/ha) or Neem oil at 3 % strength can be sprayed to kill insects.

» Some *Broscus punctatus, Liogryllus bimaculatus, Anpopus hypsipylae, Ichneumon* sp., and *Turanogonia chinensis* to take care of those pesky pests.

» *Bacillus thuringiensis* has been shown to be quite successful in the treatment of cut worms.

**Colorado potato beetle,** *Leptinotarsa decemlineata*

**Damage symptoms**

- » Allowing adults and larvae to proliferate unchecked can lead to full defoliation and practically total crop loss.

- » Adults only eat about 10 square cms of foliage every day, but larvae can consume up to 40 square cms of plant material every day.

- » Larvae consume foliage and mature plants for food.

- » The harm they do can significantly decrease yields or even kill plants.

**Pest identification**

- » The Colorado potato beetle is a little, yellow bug, about 0.5 inches long and 0.25 inches wide, with alternating black and white stripes down its back.

- » The larvae are a bright reddish orange colour and are covered in two parallel rows of black dots.

**Survival**

- » The mature Colorado potato beetle spends the winter underground.

**Management**

- » Planting different crops at different times can help reduce insect populations.

- » It is recommended to use a pesticide such as acetamiprid, azadirachtin (Neem oil), **beauvaria bassiana,** carbaryl, cyfluthrin, deltamethrin, esfenvalerate, permethrin, or spinosad.

- » Nabis damsel bugs and Geocoris big-eyed bugs are two common predators of young insects.

- » If you want to use *Bacillus thuringiensis (var. tenebrionis)* to get rid of little larvae (less than 1/4 inch), you should do so right after the eggs hatch or as soon as the larvae are spotted.

- » They found that extract from neem seeds stunted the insect's development.

**White grub,** *Holotrichia consanguinea*

**Damage symptoms**

- » For their sustenance, grubs eat tubers and roots.

- » Nighttime is when grubs do most of their eating.

## Pest identification

» "C" shaped grub describes the larval stage.

» Brown beetle with a white prothorax when fully grown.

## Management

» Mechanical control by gathering of beetles.

» Frequent ploughing before monsoon (April – May) exposes the grubs and pupae for predation by natural enemies (birds) (birds).

» Flooding of the fields for 7-10 days, wherever practicable.

» Use non-host crops in appropriate crop rotations.

» Mass data can be collected using light traps at a rate of 1/ha between the hours of 7 pm and 9 pm.

» Ten days after the first summer rain, dust with Quinalphos 5% at a rate of 25 kg per hectare.

## Epilachna beetle, *Epilachna vigintioctopunctata*

### Damage symptoms

» A harmful insect that feasts on potato leaves and eventually kills the plants.

» Whole crop failure appears to be possible after an infestation.

» Usually only one side of a leaf gets grazed, leaving the other side untouched.

» The underside of leaves is where you'll find the larvae, whereas the upper surfaces of leaves will be home to the adults.

» They eat the foliage.

### Management

» Adults and larvae must be destroyed individually. If you decide to try and get rid of the beetles in your crop, you should do it as soon as possible once you notice them.

» Weeds belonging to the Solanaceae family should be eliminated from the area.

» After harvesting, gather up any remaining crop trash and dispose of it by fire.

» The potato epilachna beetle might be deterred by ash. Don't wait till there are a lot of pests before applying to the crop.

» Natural pyrethroids may not be as effective as synthetic ones.

» The introduction of parasitic wasps (Pediobius spp.), which prey on Epilachna beetles, has resulted in their successful eradication.

# DISEASES

### Early blight, *Alternaria solani*

### Damage Symptoms

» Little, dark brown to black dots (1-2 mm) are the earliest symptoms of leaf lesions. A "target spot" lesion is characterized by the growth of a spot followed by the formation of concentric rings of elevated and depressed necrotic tissue.

» The affected leaves become brittle and fall off or turn a sickly yellow colour.

» Lesions on the tuber are often a dark colour, are irregular in shape, and have a raised, violet border.

» Dry, leathery, and typically brown, the underlying flesh is a dead giveaway.

### Favorable conditions

» Weather will be mild and dry, with brief showers.

» Low plant vitality.

» 25–30 °C.

### Survival and spread

» Primary: The pathogen survives the winter in plant waste in or on the soil, where it can do so for at least a year and possibly more. More than 17 months can pass without the conidia and mycelium of the fungus being killed off by the soil or the infected plants.

» Secondary: Spores can be carried by the wind, the water, the insects, and many animals and machines. It can also develop from seeds.

### Management

» To avoid spreading disease, only plant tubers that have not been contaminated.

» Planting Solanaceous crops near potato fields is discouraged.

» The disease can be contained by spraying a 1% Bordeaux mixture.

» Kufri Sindhuri, Kufri Jyoti, and Kufri Lalima are disease-resistant cutivars that should be used in planting.

» Maneb, Mancozeb, Chlorothalonil, and Triphenyl tin hydroxide are some of the protectant fungicides suggested for preventing the spread of late blight.

## Late blight, *Phytophthora infestans*

### Damage symptoms

» The late blight disease destroys both the potato plant and its tubers.

» Foliar symptoms consist of irregular to circular lesions on vulnerable plants.

» As one might expect, these stains are most commonly found on the very ends of leaves.

» The stems and leaf stalks often get a dark brown color as well.

» Leaves quickly become covered in spots if the weather is wet, which can be a major problem.

» Wet weather causes the upper and lower leaf surfaces, as well as the stems, to develop a white fuzz or mildew-like growth. This is evidence of spore production by the fungus.

### Favorable conditions

» Around 90% humidity or higher

» Temperatures as low as 10°C at night and a high of 25 °C during the day.

» It's going to be cloudy the next day.

» At least 0.1 mm of precipitation the next day.

### Survival and spread

» The pathogen can be maintained in the soil by means of decomposing plant matter.

» The infected tubers are largely to blame for the disease's ability to live on in subsequent harvests.

» As a result of the soil and contaminated seed tubers, it rapidly multiplies.

» The sporangia are the actual infectious agents that spread via the air.

### Management

» Seed potatoes should be confirmed as healthy.

» Kufri Sindhuri, Kufri Jyoti, Kufri Badshah, Kufri Chipsona 1 and 2, Kufri Jawahar, and Kufri Sutlej are some disease-resistant cutivars that may be grown in the North India.

» Kufri Jyoti and Kufri Giriraj in the hills of Himachal Pradesh; Kufri Kanchan and Kufri Jyoti in the hills of Darjeeling; and Kufri Swarna and Kufri Thenamalai in the hills of Nilgiri.

» Recommended cultivars include the resistant Kufri Naveen, Kufri Jeevan, Kufri Alankar, Kufri Khasi Garo, and Kufri Moti varieties.

» Spraying with one or two systemic fungicide formulations (Metalaxyl, Ridomil).

» We recommend using contact fungicides (Mancozeb, Captafol, Copper oxychloride, Bordeaux mixture) for the initial and final sprays.

## Black scurf, *Rhizoctonia solani*

### Symptoms

» Brown to black, recessed lesions on stems and stolons are typical of Rhizoctonia.

» It is possible for these cankers to continue growing and eventually girdle the stems and stolons of developing plants.

» In older plants, rhizoctonia infections rarely cause fatal girdling of the stems.

» It's important to note that these plants' health can be seriously impaired, and they may become more susceptible to other diseases, especially early blight.

» Tubers are also a target for the pathogen.

### Favorable conditions

» Excessive humidity and heat are breeding grounds for bacteria and viruses.

### Survival and spread

» The infected tubers, along with the pathogen, can be found in the soil and in plants.

» A saprophytic lifestyle on the organic matter allows the fungus to survive in the soil for years.

» The sclerotia found on seed tubers serve as the primary vector of infection for the crop that is grown from these tubers.

» Existing mycelium in the soil may grow and produce new hyphae whenever favourable conditions return.

## Management

- » To avoid spreading disease, only plant tubers that have not been contaminated.
- » Cereal crop rotation, including oats and barley.
- » Mustard and groundnut cake applied at a rate of 2.5 tonnes per hectare.
- » When it comes to preventing the spread of Rhizoctonia in tubers, the fungicide known as "Maxim" has proven to be the most effective option.
- » In addition to regulating, the combination of solar heating and treatment showed superior than solarization alone.

## Wart, *Synchytrium endobioticum*

### Symptoms

- » The only component of the plant that stays healthy below earth are the roots.
- » The centres of infection and abnormal growth activity that result in wart production are buds on stems, stolons, and tubers.
- » Some weakening of plant vitality is possible.
- » The tubers, the stems, and the stolons all develop tumours or warts.
- » The virus that causes warts spreads easily through soil and seeds.

### Favorable conditions

- » The disease tends to thrive in damp, chilly environments.

### Survival and spread

- » Resting spores of the fungus can survive in the soil for years.
- » This thick-walled winter sporangium can be discovered at depths of up to 50 cms in the soil, where they can survive for up to 30 years.
- » Infected seed is typically disseminated by the use of warted tubers.
- » Besides being conveyed on the feet of humans, animals, or farm instruments, and through manure, the wart pathogen can also be disseminated by seed of wart-resistant cultivars cultivated on infected ground.

### Management

- » It's important to plant only disease-free seeds.
- » Bean and radish should be rotated out of your garden every five years at least.

- » Throw away any tubers or plants that have been affected.
- » Burning and removing warty tubers is an effective way to get rid of plants of sensitive kinds.
- » Use the potato and maize intercropping method.
- » Grow immune potato cutivars. like Kufri Jyoti, Kufri Sherpa, Kufri Kanchan, Kufri Anand, Kufri Chipsona- 2, Kufri Frysona, Kufri Giriraj, Kufri Himsona, Kufri Jawahar, Kufri Kashigaro, and Pimpernel.

## Common scab, *Streptomyces scabies*

### Symptoms

- » Scab begins when tubers start growing.
- » Brown, roughened dots with ragged edges characterize scabs.
- » Occasionally ridged parts are in broken concentric rings.
- » Warty, sunken, or elevated spots are all possible.
- » Rusty patches, ranging from the superficial to the profound, are possible.
- » The presence of white grubs, wire worms, and millipedes in the area might grow or deepen scab marks.

### Favorable conditions

- » The low soil pH and high soil moisture conditions are ideal for the development of many diseases.
- » Overwatering can encourage the spread of disease.

### Survival and spread

- » The disease agent lives in the soil and reappears each year, contaminating the harvest.
- » The disease is transmitted mostly through the transport of infected potato tubers.
- » It's possible that the infection can survive the animals' digestive systems and spread via manure from farms.
- » The soil, uncomposted manure, and seed can all be reservoirs for pathogens.
- » The contaminated soil, seed, and water all contribute to its rapid spread.

## Management

» To avoid spreading disease, only plant tubers that have not been contaminated.

» From the time tubers begin to form until they are 1 cm in diameter, the crop should be irrigated periodically to maintain moisture at field capacity.

» Wheat, peas, oats, barley, lupin, soybeans, sorghum, and bajra should be rotated with one another and green manuring implemented to reduce the prevalence of disease.

» To maximize yield, plough potato fields in April, then leave the soil exposed during the hot months of May and June.

» Produce Patna Red, Kufri Alankar, Kufri Sindhuri, Kufri Dewa, and Kufri Lalima, all of which are tolerant to a wide range of diseases.

## Bacterial wilt, *Ralstonia solanacearum*

## Symptoms

» The wilting of an infected plant typically occurs at the plant's tips, such as the leaves or the points where the stems branch.

» The undersides of the leaves turn yellow, and the entire plant eventually withers and dies.

» A brown ring can be seen around the cut ends of the stems.

» Yet, when a tuber is split in half, dark brown or black rings become apparent. When squeezed or left alone for a long, these rings will lead a viscous white fluid.

» Fluid discharge from the tuber eyeballs is another sign. Soil that remains on the eyes of harvested tubers is one indicator of this.

» Rotting tubers are a sign of a severe infection.

## Favorable conditions

» Disease flourishes in conditions of high temperature (25 to 35°C), high soil moisture (RH > 50%), and low pH (6.2-6.6).

## Survival and spread

» The bacterial wilt pathogen is able to overwinter in the soil (host-free for multiple growing seasons), water, seed tubers, and potato plant debris.

» The disease can also spread in the field from infected seed.

» Infected seed, air, water, soil, farming tools, livestock, and people can all play a role in the rapid spread of the disease from one field to another or even inside a single field's worth of plants.

» Disease spread swiftly due to increased temperatures in warehouses and storage facilities.

## Management

» Choose tubers that haven't been contaminated with disease.

» Applying bleach powder to the soil.

» Growing non-host crops for two to three years at a time (maize, sorghum, finger millet, cereals, garlic, onion, lupin, cabbage, green manuring crops like sun hemp, mustard, etc.).

» Potato, maize, and cowpea intercropping.

» Seed tubers were soaked in a Bacillus cereus (strain B4) and Bacillus subtilis (strain B5) formulation containing 0.25% sugar for 20 minutes.

» The incidence of wilt was reduced by 76.31% and 53.15 %, respectively, when Pseudomonas fluorescens and Bacillus subtilis were applied to the tubers, while yield was improved by 220% and 129%.

» The disease incidence was reduced by 80-82.5 % and the tuber yield was enhanced from 6.04 to 6.94 tonnes per hectare (t/ha) by bacterizing the tubers and then applying soil containing an avirulent strain of Ralstonia solanacearum.

## Potato Virus Y

## Symptoms

» PVY is characterized by a wide range of symptoms, from mild mosaic to severe mosaic, veinal necrosis to leaf drop streaking, plant stunting to rugosity, bunching or twisting of leaves to down-turning of leaflet edges.

» Petiole necrosis and vein necrosis are common causes of leaf drop and stem adhesion.

» Infected plants exhibit dense, clumped crowns and either bare, hanging lower stems or necrotic leaves.

» Necrotic ring spot can be seen on infected potatoes.

» Necrotic ring spot on potato tubers.

» Reduced yields of 60-80 % are possible if PVY becomes endemic.

## Spread

» Aphids can transmit it, and it can also be carried mechanically by humans, both of which can have a significant impact on crop output.

» Aphids, such as the green peach aphid (*Myzus persicae*), are the vectors for the lyssavirus phytovinae (PVY).

» Mechanical transmission and transmission through tubers are also possible modes of transmission.

## Management

» The utilisation of virus-free seed.

» Virus in seed tubers was eradicated using a combination of thermotherapy and meristem culture.

» Use insecticides to cut down on the number of aphids.

» To prevent the spread of aphids, mineral oil is sprayed.

» Inherent resistance/tolerance to PVX, PVY, and PLRV is present in Kufri Jyoti, Kufri Kuber, Kufri Lalima, Kufri Sindhuri, and Kufri Ashoka.

## Leaf roll, *Potato Leaf Roll Virus*

## Symptoms

» When aphids attack a healthy plant during the current growing season, the plant becomes infected for the first time. The area of inoculation is the starting point for the onset of symptoms.

» Some reddening of tissue can be seen along the leaf margins and the upper leaves become pale, erect, and rolled.

» When an infected tuber is planted, it might spread the virus to other plants.

» Extreme rolling and a leathery feel can be felt on the lowest leaves.

» Plants with this condition tend to be upright and chlorotic.

» A reddening of the leaf edges or chlorosis may appear on the oldest leaves.

» Noise-making plants that rattle when disturbed.

» Internal net necrosis is visible in infected potato tubers.

## Management

» Grow from certified, high-quality seed.

» Heating the tubers at 55 °C for 15 to 20 minutes.

» Unwanted diseased plants appearing before the season even begins.

» Tolerance and resistance to PVX, PVY, and PLRV are innate characteristics of the Kufri Jyoti, Kufri Kuber, Kufri Lalima, and Kufri Sindhuri and Ashoka varieties.

## References

Secor, G. A., & Gudmestad, N. C. (1999). Managing fungal diseases of potato. Canadian Journal of Plant Pathology, 21(3), 213-221.

Frank, J. A., & Leach, S. S. (1980). Comparison of tuberborne and soilborne inoculum in the Rhizoctonia disease of potato. *Phytopathology*, *70*(5), 1-53.

Kreuze, J. F., Souza-Dias, J. A. C., Jeevalatha, A., Figueira, A. R., Valkonen, J. P. T., & Jones, R. A. C. (2020). Viral diseases in potato. *The potato crop: its agricultural, nutritional and social contribution to humankind*, 389-430.

Irshad, G., & Naz, M. F. A. F. (2014). Important fungal diseases of potato and their management–a brief review. *Mycopath*, *11*(1).

Gudmestad, N. C., Taylor, R. J., & Pasche, J. S. (2007). Management of soilborne diseases of potato. *Australasian Plant Pathology*, *36*, 109-115.

Giordanengo, P., Vincent, C., & Alyokhin, A. (2013). Insect pests of potato. *Glob. Perspect. Biol. Manag. Elsevier Inc Wyman Str. Walth. USA*.

Alyokhin, A., Rondon, S. I., & Gao, Y. (Eds.). (2022). *Insect pests of potato: global perspectives on biology and management*. Academic Press.

Kroschel, J., & Schaub, B. (2013). Biology and ecology of potato tuber moths as major pests of potato. *Insect pests of potato. Global perspectives on biology and management*, 165-192.

Gao, Y., Alyokhin, A., Nauen, R., Guedes, R. N., & Palli, S. R. (2022). Challenges and opportunities in managing pests of potato. *Pest Manag Sci*, *78*(9), 3729-3730.

Chandel, R. S., Chandla, V. K., Verma, K. S., & Pathania, M. (2022). Insect pests of potato in India: biology and management. In *Insect pests of potato* (pp. 371-400). Academic Press.

# 20

# Pumpkin

## INSECT PESTS

### Fruit fly, *Zeugodacus* (*Bactrocera*) *cucurbitae*

#### Damage symptoms

» Maggots cause decay in immature and ripe fruits, as well as drying and shrinkage in immature fruits.

» Because of the presence of fruit flies, the affected fruits leak a resinous fluid and take on an abnormal appearance. Maggots cause fruit to rot and fall off trees before its time because they eat the fruit pulp.

» If the plant is under threat, it might wilt all the way down to its roots.

#### Host range

» Cucumbers, melons, snake melons, chow chows, pumpkins, bitter melon, water melon, and coccinia are all examples of unusual vegetables.

#### Favorable conditions

» The prevalence is higher in damp environments.

#### Management

» Don't just let rotting or otherwise damaged fruit go to waste; instead, gather it up and burn it in large pits.

» Ribbed gourds can be used as a trap crop.

» Citronella oil, eucalyptus oil, vinegar (acetic acid), and lactic acid are all effective fly attractants.

» Spraying Neem soap at 1% or Pulverized NSPE at 4% at 10-day intervals after flowering and applying Neem cake at 250 kg/ha to the soil right after germination and again at flowering is recommended.

» 10 ml of the bait (a mixture of Methyl eugenol and Malathion 50 EC) should be stored in polythene bags at a rate of 25 ml per hectare.

» For every acre of land, set up three cue lures (para pheromone traps) to entice and catch male fruit flies.

» Carbaryl 50 WP at 3 g/L or Indoxacarb at 0.5 ml/L sprayed.

» Fruit fly infestation in bitter gourd was reduced after treatment with 0.05% acephate and 1 ml of a commercial preparation of *Beauveria bassiana* per litre of solution (Maicykutty and Gopalakrishnan, 2003).

» Effective natural enemies include *Pediobius foveolatus* and *Tetrastichus ovularum*.

» Using measures to increase resistance, such as using resistant plant kinds, etc. Arka Suryamukhi (pumpkin).

## Damage symptoms

» Mature pumpkin beetles eat the leaves of cucurbits, leaving behind only the veins and occasionally larger holes.

» Even a tiny population of beetles can completely defoliate and kill a young plant.

» Seedling stunting; leaf, stem, and petiole damage; plant stand reduction; plants may show scars on fruit from beetle feeding damage.

» Older plants can take more abuse before they begin to produce less fruit.

» Root damage by larvae is usually minor, and plants suffer only occasionally.

» The bugs that cause pumpkin damage typically cluster together on either the young or the old leaves of the crop.

## Pest identification

» Larvae of the striped and spotted cucumber beetle are white or yellow overall, but have darkened head capsules and an elongated dark patch at the very end of the abdomen. The larvae of the spotted cucumber beetle are more elongated than those of the striped kind.

» Adult striped cucumber beetles have a bright yellow colour and have three black stripes running lengthwise across their backs. In length, they measure around 0.25 inches, and their bellies are black. The abdomen of an adult spotted cucumber beetle is yellow, and the rest of its body is either yellow or green with 12 black spots.

## Host range

» Melons, pumpkins, cucumbers, and squashes (like zucchini) make up the majority of this group.

## Survival

» Beetles spend the winter underground in places like dirt and leaf litter, emerging when the temperature rises above 12.7 °C.

## Management

» Check for beetle activity frequently in newly planted areas.

» In areas where the pest problem is mild, it may be possible to gather and kill the insects mechanically.

» Protecting plants with floating row coverings is possible, but they must be removed as flowers form so that pollinators may pollinate the crop.

» Kaolin clay applications for controlling tiny beetle populations have shown promise.

» Spray Indoxacarb 14.5 SC at 0.5 ml/L, Carbaryl 50 WP at 4 g/L, Quinalphos 25 EC at 2 ml/L, Chlorpyrifos 20 EC at 2.5 ml/L, or Trichlorfon if the incidence rate is high.

### Serpentine leaf miner, *Liriomyza trifolii*

### Damage symptoms

» Pumpkins, cucumbers, and other squash show particularly high rates.

» While other vegetables may succumb, bitter gourd appears to be resilient.

» Tunnels made by larvae mining between the upper and lower leaf surfaces are twisting and white at first, but gradually expand as the larvae develop.

» If the leaves dry up, the fruit can get sunburned and the crop's productivity and quality suffer.

» Leaf mining can lower yields or even kill plants in extreme infestations.

» Late in the season is when you'll see the worst infestations.

**Pest identification**

- » Larvae are around 2 mm in length and look like small, bright yellow maggots.
- » Little brown grains of rice best describe the appearance of pupae.
- » Adults of this species are tiny (1.5 mm), black and yellow flies with a bright yellow scutellum, a triangular patch on the top back of the thorax.

**Host range**

- » Cucumber, pumpkin, and watermelon.

**Management**

- » Infestations of harmful leaf miners, which often begin around the middle of July, are a good reason to postpone summer plantings.
- » Use row covers (plastic and spun-bonded materials) before planting to prevent leaf miners, then remove them gradually as flowers appear, if necessary.
- » An effective strategy for controlling leaf miner is to remove infected cotyledon leaves one week after germination, then spray with Neem seed powder extract at 4% or Neem soap at 1%.
- » *Hemiptarsenus vericornis*, a natural larval parasitoid, has been reported to be the most common parasitoid on this pest.
- » *Diglyphus* species. and *Chrysocharis* spp. may be the most crucial parasitoids for preventing leaf miner infestations.
- » Apply 250 kg per hectare of Neem cake to the soil after planting.
- » If the rate of infection is high, you should start by destroying any badly affected leaves. Next, spray a mixture of 5 grammes of Neem soap and 1 millilitre of Hostothion per litre. After a week, you can spray Neem soap at 1% concentration or NSPE powder at 10000 ppm or higher (2 ml /L).

## Aphid, *Aphis gossypii*

**Damage symptoms**

- » While melon aphids prefer feeding on the undersides of leaves, they have been known to target the tender new growth as well.
- » They drain the plant's vital fluids, leading to distorted leaves and the untimely demise of growing tips.
- » Honeydew, a sticky substance left by melon aphids, can encourage the growth of sooty mould on the leaves and developing tips.

» Being the aphids' caretakers and eaters of honeydew, ants are sometimes spotted in areas where aphids are prevalent.

» Melon aphids tend to congregate, so you can find a lot of them on a small number of leaves.

» When the damage is severe, the crop may be lost.

» Most importantly, melon aphids are the primary vector of the mosaic virus, which further exacerbates the harm they inflict.

## Host range

» Marrow, melon, and cucumber.

## Management

» Infestations of aphids can be managed through selective pruning if they are confined to a small number of leaves or shoots.

» Before planting, make sure the transplants are free of aphids.

» Tolerant cultivars should be used if they are available.

» Aphids can be discouraged from feeding on plants by using reflective mulches, such as silver coloured plastic.

» Spraying a forceful jet of water onto sturdy plants helps dislodge aphids from leaves.

» Spraying insecticidal soaps or oils like Neem or Canola oil is a common and effective technique of elimination.

» Potassium soap, dimethoate, maldison, petroleum oil, or primicarb are all effective pesticides for aphids.

» Pymetrozine (Fulfill), Imidacloprid (Admire), and Thiamethoxam are examples of pesticides with a reduced danger to humans (Platinum or Actara).

» Effective control may be provided all summer by naturally occurring populations of the convergent lady beetle, *Hippodamia convergens*.

» Parasitic wasps like the *Aphidius*, *Diaeretiella* and *Aphelinus* species, as well as more general predators like lacewings and syrphid larvae, prey on aphids as well.

» Aphids are susceptible to the fungal disease *Beauveria bassiana*. Effective use of *B. bassiana* necessitates three applications spaced out every five to seven days.

**Melon thrips, *Thrips palmi***

**Damage symptoms**

- » Plants ravaged by melon thrips are a lost cause.
- » Both nymphs and adults are suspected of being vectors for the tospo virus, which they spread by feeding on the developing ends of shoots and various areas of flowers.
- » The leaves turn different shades of brown, yellow, or white before wilting and falling off.
- » Fields with a lot of pests can occasionally take on a bronze color.
- » There may be discoloration, retardation, and deformation of the terminal growth if it has been damaged.
- » In the warmer months, they mean business.

**Pest identification**

- » Eggs range from colourless to a very light white, and they have the appearance of a bean.
- » While tiny and without wings, larvae are structurally similar to their larger, adult counterparts.
- » Both the pre- and post-pupae look like adults and larvae, except with the addition of wing pads.
- » The bodies of adults are a drab pale yellow or white, however they are covered in a dense covering of dark setae. The intersection of the wings creates a dark slash along the middle of the back.

**Host range**

- » Those affected include watermelon, cantaloupe, cucumber, melon, pumpkin, and squash (zucchini).

**Survival**

- » Thrips survive the cold season by hibernating in plant matter or on weeds like winter annuals that grow in or near agricultural areas.

**Management**

- » Use Neem cake to the soil twice, once just after germination and again just before flowering, then alternate applications of NSPE at 4% and Neem soap at 1% every 10-14 days.

» Any systemic insecticide, such as Acephate 75 SP at 1 g/L or Dimethoate 30 EC at 2 ml/L, can be sprayed.

» The predatory thrips *Franklinothrips vespiformis* and the tiny pirate bug *Orius insidiosus* were among the most notable predators found.

» Melon thrips are susceptible to a variety of fungi, including *Beauveria bassiana, Verticillium lecanii,* and *Hirsutella* sp.

## Leaf eating caterpillar, *Diaphania (=Margaronia) indica*

### Damage symptoms

» The caterpillars eat the young, succulent leaves, but also the leaves, soft stalks, and fruit of many other cucurbits.

» Large portions of plants can be destroyed, and the leaves can be bound together with silk and faeces.

» On rare occasions, the larvae have been found to cause damage to the blossoms.

» Later in development, when the fruit is in close touch with the ground, the larvae can eat small holes in the surface.

» These flaws are a result of the damage that this produces, and they lower the quality of the fruit.

### Pest identification

» There are 1-2 mm of length to each egg that is a creamy white tint.

» Their long, shiny bodies range in colour from light to dark green, and they have a white stripe along either side. The full-grown length of a larva is about 15 mm.

» Mature moths typically have a wing span of 20-25 mm. There is a dark brown stripe at the edge of the white wings.

### Host range

» To name a few: watermelons, cantaloupes, cucumbers, and pumpkins are their favourites.

### Management

» The smallest larvae should be targeted for elimination because they are the easiest to eradicate.

» Use Neem cake to the soil twice, once just after germination and again just before flowering, then alternate applications of NSPE at 4% and Neem soap at 1% every 10-14 days.

» This pest can also be effectively controlled by spraying Neem or Pongamia soap at a concentration of 0.75 %.

» The only organic chemical that exists is Entrust.

» The chemicals SpinTor, Spinatoram, Intrepid, and Avaunt are among those with reduced risk.

» Besides Lannate, other compounds include the pyrethroids (Brigade, Asana, and Warrior).

» Use any contact pesticide, such as Carbaryl 50 WP at 3 g/L, Indoxacarb at 0.5 ml/L, or Trichlorfon.

» *Bacillus thuringiensis* (Dipel) and (*Trichgramma chilonis* + *Dolichogenidea stantoni*) were shown to be the most efficient biocontrol agents.

» *Bacillus thuringiensis* is the most effective biological control agent for cucumber moth. It's more eco-friendly yet takes longer to kill the larvae.

## Red pumpkin beetle, *Aulacophora foveicollis, A. cincta, A. intermedia*

### Damage symptoms

» Mature beetles make uneven holes in the lamina of leaves as they eat.

» When the crop is at the cotyledon stage, the greatest amount of harm is done. Hence, the initial generation causes more harm than later generations.

» The adult insects feed on mature plant leaves, removing chlorophyll to create a net- like pattern on the foliage.

» In extreme circumstances, resowing the crop may be required because the injured plants have withered away.

» The larvae wreak havoc by eating soil-contact leaves and fruits as well as burrowing into the roots and underground stem section.

» Roots and subterranean stem sections that have been injured may decay if infected by saprophytic fungus.

» Infested creepers may cause the immature and smaller fruits to dry up, while the larger, mature fruits become unfit for human eating.

## Pest identification

» Newly born grubs are a dingy white tint, whereas full-grown grubs are a creamy yellow.

» For adults, you can choose between the red *Aulacophora foveicollis*, the grey *Aulacophora cincta* with its iridescent yellow-red border, or the blue *Aulacophora intermedia*.

## Host range

» Besides bitter gourd, pumpkins, bottle gourds, cucumbers, snake gourds, melons, and coccinia.

## Management

» As soon as you're done harvesting, plough up the fields and kill off any adults that may have gone into hibernation.

» Kill or capture adult beetles.

» Spray either 50 EC of malathion or 30 EC of dimethoate or 25 EC of methyl demeton at a rate of 500 ml/ha.

» It has been found that spraying with contact insecticides, such as Carbaryl 10% DP, will effectively reduce beetles, although the treatments must be repeated at 7-day intervals.

» It has been found that incorporating Neem oil cake into the soil can eliminate bug larvae.

» Plant seeds early so that your plant has already emerged from its cotyledon stage before the beetles start feeding.

## Epilachna beetle, *Henosepilachna vigintioctopunctata*

## Damage symptoms

» A large population of grubs, including juveniles and adults, may be seen on the undersides of the leaves.

» Both the adult moth and the grub stage feed on the tops of leaves, skeletonizing them so that they seem like delicate lace.

» When leaves are assaulted, they become brown, dry out, and fall off.

» There is a very sickly appearance to the crop if the infection is severe.

» Negative effects on the plant's vitality and productivity can be expected.

**Pest identification**

- » The adults of this species are between 6 and 10 mm in length and have 28 black spots over their white, orange, or red bodies.
- » Yellowish larvae with black or brown spines that branch off are typica

**Management**

- » To prevent population growth in the intervening time between harvests, it is best to clear away crop remains as soon as possible.
- » Use of row coverings helps keep beetles out of crops and reduce damage.
- » Picking and killing eggs, grubs, and adults on a regular basis can successfully reduce the pest population and stop an infestation in its tracks if the cropped area is relatively small.
- » In the morning, you can shake the larvae and adults down into a container of kerosene.
- » Spraying the pest with Malathion, Fyfanon, or Zythiol 50 EC at a concentration of 2 milligrammes per millilitre of water will effectively eliminate it. The treatment needs to be applied as soon as possible after the pest is spotted, and then repeated every 15 days.
- » Use pesticides like Carbaryl (Sevin) or pyrethroids to spray the area.
- » Controlling larvae with Azadirachtin or Spinosad.
- » Parasitoids such as the Braconid wasp and the *Celatoria setosa* are a real pain (Tachinid fly).

**Two-spotted spider mite, *Tetranychus urticae***

**Damage symptoms**

- » During the dry season, when temperatures are high and humidity is low, damage typically occurs.
- » The mites pierce and suckle the underside of the leaves for food.
- » When a leaf is damaged, the veins and midrib become paler or bronzed first.
- » Mottled, silvery-yellow leaves may drop off the plant early if an infection is bad enough.
- » If the incidence is high, you can see colonies of mites within the silk web on the underside of the leaf.

» The quantity and quality of fruit produced may suffer if pest infestations are severe.

## Pest identification

» Oval and slightly see-through, eggs are a common household item.

» Mature mites measure around 0.06 inches in length, have four legs, range in colour from bluish-green to pinkish-cream, and are marked with two dark dots. They can turn an orange/red color under unfavourable or cold weather.

## Host range

» Melons and other gourds are particularly popular.

## Management

» The use of 1% Neem or Pongamia soap in a spray bottle. Cover the bottom with spray.

» If you prefer, you can use a spray formulation of Dimethoate 30 EC at 2 ml/L, Ethion 50 EC at 1 ml/L, or Wettable Sulfur 80 WP at 3 g/L instead.

» Potassium soap, the miticide Dicofol, or a combination of Dicofol and Tetradifon can be sprayed to kill the insects. Unlike Dicofol, which only affects adults and nymphs, Tetradifon is effective against both egg stages.

» It is possible to effectively manage red spider mites on cucurbits by releasing predatory mites, such as Phytoseilus persimilis. At the earliest sign of a mite infestation, you should release ten thousand predators across 200 square metres of crop.

» As for other predators, there's the western predatory mite (*Galendromus* [*Metaseiulus*] *occidentalis*), six-spotted thrips (*Scolothrips sexmaculatus*), western flower thrips (*Frankliniella occidentalis*), lady beetles (*Stethorus* etc.), minute pirate bug (*Orius tristicolor*), and lacewing larvae (*Chrysoperla carnea*).

# DISEASES

## Downy mildew, *Pseudoperonospora cubensis*

## Symptoms

» Mottled yellow angular patches emerge on the upper surface of the leaves, bounded by the veins, giving the appearance of a mosaic pattern.

» As it rains, the underside of these patches gets a velvety purple coating.

» In age, the spots develop a necrotic appearance.

» Infected leaves turn yellow, wither, and fall off.

» Infected plants wither and eventually die.

» The quality and maturity of the resulting fruit may suffer.

## Host range

» This delete impacts a wide variety of vegetables.

## Favorable conditions

» Fog, heavy dew, and high humidity all contribute to the spread of the disease.

» Humidity and temperature are particularly conducive to the spread of this disease.

» Above 90% relative humidity.

» Very wet ground.

» Normal rainfall pattern being interrupted frequently.

## Survival and spread

» Originating from soil oospores and sporangia of nearby perennial collateral weed hosts.

» Secondary: conidia (sporangia) or free-floating zoospores that are carried by the wind or rained upon.

## Management

» Pick off any leaves that look sick and throw them away.

» The use of a bed system with ample space, adequate drainage, air flow, and sun exposure aids in preventing the spread of disease.

» If you want your plants to thrive, you need to give them space.

» Don't water the lawn from above.

» From a central location, water the plants.

» At 7-day intervals, spray either Ridomil (0.3%), Blitox (0.2%), or Mancozeb (0.2%).

» For effective disease management, apply Dithane M-45 (0.3% every 15 days).

» *Bacillus subtilis*, a biocontrol agent, is available as a wettable powder that can be applied to plants to prevent downy mildew.

## Powdery mildew, *Erysiphe cichoracearum*, *Sphaerotheca fuligenia*

### Symptoms

» Infected leaves and stems will have patches ranging in colour from mealy white to filthy grey.

» The plant's green surface becomes covered in a powdery substance that accelerates the wilting, drying, and defoliation process.

» In the case of a serious infection, the fruit may also develop a white powdery material on their surface.

» Baby fruits lose their colour, stay immature and misshapen, and eventually fall off.

» The fruit's quality declines as a result.

### Host range

» Fall squash, water melons, bottle gourds, coccinia, cucumbers, and ridge melons are all vulnerable. The bitter gourd seems to be somewhat immune.

### Favorable conditions

» Temperatures in the 60s and lots of shade are ideal for the spread of this disease.

» Temperatures above 35 °C and relative humidities above 70 % are ideal for the spread of disease.

» High morning humidity (above 90%).

» Season of cool, dry weather.

### Survival and spread

» Conidia or latent mycelium from diseased plant waste or a secondary host will ensure your survival.

» Spreading from one plant to another, the spores ride on the wind currents.

### Management

» Elimination of pathogen hosts, including sick plants and weeds.

» Penconazole 0.25%, Tridemorph 1%, Benomyl 0.15 %, Karathane 0.05 %, Hexconazole 1%, Elosal 0.5 %, wet 2% sulphur, and carbendazim 0.1% spray.

» As a form of biological control, the parasitic fungus *Ampelomyces quisqualis* spores are used to eradicate the powdery mildew. *Sporothrix flocculosa* (syn. *Pseudozyma flocculosa*) and *Bacillus subtilis* (a kind of bacteria) also showed promise.

» To prevent muskmelon cutivars from rotting, use a variety that is resistant to the disease. Arka Rajhans, Florigold, Gulfstream, Jacumba, Edisto, and Cantaloupe No. 45 here.

## Anthracnose, *Collectorichum lagenarium*

### Symptoms

» Leaves develop small yellow or water-soaked spots, which quickly expand and turn brown in most cucurbits but black in water melon.

» Symptoms include peeling and flaking bark or complete leaf dehydration.

» After the stem becomes infected, the entire vine will perish.

» Infection in the fruit pedicel can cause the developing fruit to turn black, wither, and dry.

» Some fruits develop black, round cankers.

» Depending on the host and the surrounding environment, the patches can be as large as 5 cms in diameter.

» Dark, depressed areas often have spores of a salmon colour.

### Host range

» Produce like a cucumber, watermelon, and chow chow.

### Favorable conditions

» Conidia do not germinate below 4.4°C or over 30°C or in non-humid circumstances, hence high humidity and temperature of 24°C is optimal for disease development.

### Survival and spread

» The fungus can survive the winter on infected cucurbit crop debris and be transmitted to cucurbit seeds.

» The conidia (airborne spores) are released by the fungus in the spring when the weather is damp, infecting the cucurbit vines' leaves.

## Management

» Seeds must be disease-free and verified.

» A treatment rate of 3 g/kg seed is recommended when using Thiram or Captan on seeds.

» Securing and eradicating diseased vegetation.

» Get rid of and burn the remnants of the old cucurbit vines.

» Change up your cropping method annually.

» Spraying with 0.2% solutions of Difolatan, Dithane M-45, or Bavistin has proven helpful.

» *Bacillus subtilis* (QST713) has been suggested as a potential biological control agent. Strong promise exists for the use of an endophytic *Streptomyces* sp. (strain MBCu- 56) in the management of cucumber anthracnose.

## Gummy stem blight, *Didymella bryoniae*

## Symptoms

» Between the leaf veins, grayish-green lesions form.

» Stems develop tan or grey spots.

» Leaves develop a vague necrosis at their margins, followed by huge, wedge-shaped necrotic patches.

» Water-soaked sores emerge first on infected stems, followed by a tan coloration. Gum-like beads of a rusty brown or black colour may emerge from a stem lesion.

## Host range

» Cucumber canker, pumpkin blight, and watermelon wilt are all devastating diseases.

## Favorable conditions

» Diseases spread most effectively between 20 and 24 °C with plenty of humidity in the air.

» The fungus is at its most destructive in hot, muggy conditions.

## Survival and spread

» Without a host, the fungus can live for two years in crop leftovers and soil.

» Conidia spread from the original source to the newly planted crops via splashing rain, irrigation water, and winds.

## Management

» Choose a dependable source for seeds that have been treated to prevent disease.

» A treatment rate of 3 g/kg seed is recommended when using Thiram or Captan on seeds.

» Plants that are not hosts for pests should be rotated in regularly.

» No overhead watering, please.

» At the end of the growing season, pick through your vines and fruit trees and get rid of any that look sick.

» As soon as the sickness is detected, spray Bavistin (0.2% concentration). Spraying with Dithane M-45 (0.25%) or Propiconazole (0.1%) may be necessary if the disease is not under control.

» Fungicide use could be reduced with the use of chlorothalonil (Bravo Ultrex 82.5 WDG at 3.0 kg/ha), *Bacillus subtilis* (Serenade 10WP at 4.5 kg/ha), and pyraclostrobin plus boscalid (Pristine 38 WG at 1.0 kg/ha).

» Use of synthetic fungicides was reduced when biofungicides were rotated with Chlorothalonil (Zhou and Everts 2008).

» Fruit should not be damaged in any way before or after harvest, as this can allow pathogen entry when it is being stored.

## Alternaria leaf blight, *Alternaria cucumerina, A. alternata*

## Symptoms

» Little brown dots that have a halo of yellow or green develop initially on the oldest leaves and spread quickly.

» Large necrotic patches, sometimes arranged in a concentric pattern, develop from lesions as the disease advances.

» The lesions join together, the leaves curl, and the plant dies.

» Fruit decay is another effect of the fungus.

## Host range

» Vegetable marrow, snake gourd, cucumber, watermelon, and muskmelon all sport this pattern.

## Favorable conditions

» It grows best in areas with high humidity (frequent rainfall) and temperatures between 15 and 32 °C.

## Management

» Plant only disease-free seeds.

» Keep rotating non-cucurbitaceous crops every three years.

» Seeds treated with Captan at a rate of 3 g/kg.

» Maintain clean conditions by removing agricultural trash after harvest.

» Eliminate any weeds or volunteer cucurbit plants that could be harbouring disease spores.

» Reduce leaf moisture, which promotes the growth and spread of disease, by watering plants from below rather than above.

» Before the disease may spread, spray it with Bavistin (0.1%). Treatment with Miltox (0.2%) or Dithane M-45 (0.2%) every 15 days is successful for management.

» Plant resistant hybrids and cultivars.

## Cercospora leaf spot, *Cercospora citrullina*

### Symptoms

» Little water stains or yellow flecks appear first on leaves.

» Rapid enlargement causes spots to change shape from round to irregular, with white, tan, or pale brown in the middle and a dark purple or nearly black border.

» Several spots join together to form a single, larger spot.

» In extreme cases, the leaf may dry out and die, giving the impression that it has been burnt.

» Leaves with a severe infection fall off.

» Attacks on fruits are rare but can cause them to shrink.

### Host range

» Fruits and vegetables such as watermelons, muskmelons, and cucumbers are affected.

## Survival and spread

»   This pathogen overwinters on weeds growing in the spaces between cucurbit crops and on diseased crop debris.

»   The infectious spores (conidia) of this disease are easily dispersed by the wind, splashing rain, and irrigation water.

## Management

»   Planting cucurbitaceous crops one after the other is not recommended.

»   Don't use an overhead sprinkler and water first thing in the morning.

»   Remove old crop residues that could spread disease and any unwanted cucurbitaceous weeds to maintain sanitary conditions.

»   In order to avoid the spread of disease, infected plants must be removed and destroyed.

»   Right after the last harvest, you can also do some deep ploughing.

»   Dithane M-45 at a concentration of 0.2% can be sprayed on the crop.

## Fusarium wilt, *Fusarium oxysporum*

## Symptoms

»   Infected seedlings quickly rot and perish in the moist soil.

»   In damp conditions, decaying stems may develop a pinkish or white fungal growth.

»   The leaves begin to turn yellow at the plant's roots and work their way up the stem.

»   Sometimes a single side of a leaf, a single branch, or even the entire plant will begin to wilt or turn yellow.

»   The wilting of yellow leaves is obvious before they fall off.

»   Midday, when sunlight is strong and temperatures are high, is when wilting is most likely to occur.

»   Both the size and quantity of the fruits produced by infected plants suffer.

## Host range

»   Cucumber (*Cucumis sativus*), Watermelon (*Citrullus lanatus*), and Muscaria (F. o. f. sp. *melonis*) (F. o. f. sp. *cucumerinum*).

## Favorable conditions

» Light, sandy, slightly acidic (pH 5-5.5), and low-nitrogen (particularly ammoniacal) soils are ideal for disease development.

» Dry conditions and soil temperatures between 20 and 30 °C are ideal for the disease to flourish.

## Survival and spread

» Causing fungi can be found in decomposing plant matter, other host plants, seed, or soil.

» Transferring infected dirt throughout a field is one way it might spread, while moving infected machinery or planting infected seeds are two other ways it can spread.

## Management

» Infectious soil pathogens can be reduced through solarization.

» The use of virus-free seed stock.

» Alternatively, you can soak the seeds in hot water (52°C for 30 minutes) or treat them with Benlate/Bavistin (2 g/kg seed).

» Infected vegetation is collected and destroyed by fire.

» Garlic, radishes, onions, and beets should be rotated every four years.

» *Trichoderma viride* or *Pseudomonas fluorescens* seed treatment and soil application.

» AMF *Gigaspora margarita* used in conjunction with a charcoal composting technique ranging from 3-15%.

» Fungi that aren't harmful, like Fusaria, mixed with glowing bacteria.

» Replace weaker rootstocks with hardier ones, such as *Lagenaria siceraria* or a *Cucurbita, moschata C. maxima* hybrid.

## Phytophthora blight, *Phytophthora capsici*

## Symptoms

» The leaves develop large, uneven brown blotches.

» Water-soaked, irregular lesions on the stems and leaf petioles range in colour from pale to dark brown.

» Root and crown rot cause leaves to wilt and the plant to eventually die.

» Blackened crowns and roots indicate infection.

» As a result of root death, plants are simple to remove out of the ground.

» The decay of the fruit is delicate and wet.

» The initial site of infection could be anywhere on the fruit, including the point of contact with the earth, the stem's attachment to the fruit, or a random, circular area.

» Infected fruit is mushy and gives easily to puncturing.

» Infected fruit is covered in a white, powdery fungus growth.

## Host range

» Vegetables such as cucumbers, muskmelons, pumpkins, and watermelons.

## Survival and spread

» Phytophthora can survive the winter in soil and decaying plant matter.

» Zoospores facilitate this spread, as they are able to traverse wet films and saturate soils in search of a new host plant.

» Windblown rain or soil clinging to tools used in an infected region might spread the spores from one location to another.

## Management

» Choose grounds that have good drainage.

» To facilitate water runoff, construct elevated beds.

» Stay away from growing squash, cucumbers, and tomatoes (Solanaceae) for at least three years.

» If you have a tiny garden, remove and kill any sick fruit or vines.

» Until 5 weeks after sowing, cucurbit seedlings can be protected with Mefenoxam (Apron XL LS at 0.42 ml per kg of seed) or Metalaxyl (Allegiance FL at 0.98 ml per kg of seed).

» Weekly applications of copper sulphate (i.e., Cuprofix Disperss at 2.25 kg/ha) and dimethomorph (Acrobat 50WP at 448 g/ha).

» The use of Acrobat and copper in conjunction with the seed treatment Apron 50WP.

**Angular leaf spot,** *Pseudomonas syringae pv. lachrymans*

## Symptoms

> » First, they show up as watery lesions held together by the leaf's angular veins.

> » Sometimes there is a halo of yellow around a lesion.

> » Lesions can grow from yellow to brown and eventually tear leaf tissue.

> » If the weather is moist, bacteria could flow out of places and then dry into a white residue.

> » When fruit is infected, a brown, round, superficial, firm rot quickly distorts it.

> » Flesh may also rot from the inside out.

## Host range

> » Cucumbers are the most common target, although other fruits, such as melons, squashes, and pumpkins, could also fall victim.

## Favorable conditions

> » Temperatures between 24 and 28 °C, with considerable humidity, are optimal for the pathogen's growth.

## Survival and spread

> » Bacteria can survive for up to 2.5 years in crop debris, where they can overwinter during the winter.

> » The disease can be spread through infected seed, splashing rain, irrigation water, soil, insects, farm instruments, and human hands.

## Management

> » Do summertime thorough ploughing and solarization to prepare the soil for fall planting.

> » Don't plant anything unless you're sure the seed is disease-free.

> » Seeds are soaked in hot water (50 °C) to speed up the growth process.

> » Keep the field clean and tidy while the crops are growing, and clear it of all trash as soon as harvesting is complete.

> » Crops other than those in the cucurbitaceae family should be rotated in every 3 years at the very least.

> » As possible, avoid watering from above, as splashes and dripping water

might spread bacteria.

» For example, at a concentration of 400 parts per million (ppm), streptocycline is highly efficient against bacterial plant diseases.

» Reducing the severity of the disease frequently requires the use of copper-based bactericides at intervals of 4-7 days.

» Prior to the onset of symptoms, copper compounds can be administered every 5-10 days in disease-friendly conditions.

» Insecticides are sometimes used to get rid of pesky insects, but they can harm beneficial insects and leave sores that bacteria can utilise to get in.

» Predator-Controlled Organism The *Pseudomonas geniculata* L33 strain improved crop health and production.

## Bacterial wilt, *Erwinia tracheiphila*

### Crop losses

» Crop losses of up to 75% are possible.

### Symptoms

» Clear chlorosis (yellowing) between the veins, while the major veins in the leaf maintain a dark green.

» Necrotic interveinal tissue develops throughout time (brown in color).

» There is necrosis on the leaf margins and the plant's leaves become shorter and tuft together.

» The leaves eventually turn brown and yellow around the edges and eventually fall off.

» If the stem is severed and the two ends are progressively pulled apart, strings of bacterial exudate will leak out, proving the presence of disease.

» A few weak vines may eventually fall victim to the disease.

» There is a sickness that travels via cucumber beetles.

### Survival and spread

» Only survives the winter in the digestive tracts of striped cucumber beetles (where it may make up between 1 and 10 % of the population).

» The striped and spotted cucumber beetle spreads the bacterium through its faeces when it begins feeding in the spring.

## Management

- » Avert disaster by picking resistant plants.
- » Inspect your cucumber plants once a week for wilt and cucumber beetles.
- » Reduce the number of cucumber beetles attacking your plants.
- » Kill mature beetles by picking them out one by one.
- » Application of effective pesticides to the soil and the leaves of plants may be useful for population management.
- » If beetle populations are high enough to cause wilt and the variety is particularly vulnerable, foliar pesticides may be required.

## Watermelon bud necrosis disease, *Watermelon Bud Necrosis Virus* (WBNV)

### Symptoms

- » Leaves will first show signs of minor wrinkling and chlorotic mottling, as well as yellow spots or patches.
- » In turn, vines die due to bud necrosis at their developing tips.
- » Young plants suffer extreme wilting and dieback, resulting in a total loss of the yield.
- » Shortened internodes, the upright development of younger shoots, and necrosis of the stem, petiole, and fruit stalk are all symptoms of a crop that has reached maturity.
- » Little, misshapen, and often unmarketable fruits with necrotic or chlorotic rings are produced by infected plants.

### Host range

- » WBNV is restricted to cucurbitaceous hosts, such as the ridge gourd and the cucumber.

### Favorable conditions

- » The disease tends to spread more quickly during dry, hot spells when the number of thrips is high.

### Survival and spread

- » Several different types of thrips are responsible for the spread of tospoviruses from plant to plant. *T. palmi* may be a vector for WBNV.

» Weeds and ornamental plants may play a role in the transmission and survival of these viruses by acting as reservoirs.

» WBNV can spread across Cucurbitaceae plants through their sap.

## Management

» Screening mesh (72 mesh/192 microns) and field sanitation aid in reducing WBNV in nurseries.

» Insecticides for controlling thrips.

» Invasive plant management; removing volunteer cucurbits.

## Cucumber mosaic disease, *Cucumber Mosaic Virus* (CMV)

### Symptoms

» When an infection starts between the sixth and eighth leaf, it manifests on the youngest, most rapidly growing leaves.

» Upon maturation, immature leaves develop the characteristic mosaic pattern.

» Curling leaves that are also twisted, wrinkled, and smaller than normal are all symptoms of this disease.

» Veins look lumpy because their internodes have shrunk.

» Even if an infection strikes in the middle of the season, the crop that has already been established will continue to mature normally.

» If an infection strikes a crop when it is still young, it will have a very low fruit set.

» Misshaped, mottled, warty, and shrunken fruits are common.

### Host range

» The disease destroys cucumber, pumpkin, and gourd crops.

### Survival and spread

» Several perennial weeds, which aphids find particularly appealing in the spring, serve as reservoirs for the virus over the winter.

» The aphids *Aphis craccivora* and *Myzus persicae* serve as vectors for the spread of CMV, although the insect can also infect seeds and a variety of weeds.

» Workers who pick fruit may be exposed to the disease if they touch any fruit or become contaminated with sap.

## Management

» Elimination of the host weeds.

» Occasionally, plants will become infected with rouge.

» Keep pests out of your cucumber field by planting a barrier crop like sunflower, sorghum, or pearl millet around your crop all season long.

» To prevent aphids from spreading, cover soil with a layer of mulch made of sawdust, wood chips, or aluminium foil.

» Aphid vectors can be managed by a weekly spraying of 0.05% Dimethoate, 0.05% Monocrotophos, 0.03% Confidor, or 0.02% Metasystx.

» Mineral oil sprays between 0.75 and 1.0% should be applied once a week.

» The aphid vector population can be effectively controlled by using parasitoids such as *Aphidius colemani, A. matricariae, Lysiphlebus fabarum,* and *Binodoxys angelicae.*

## References

Khan, M. M. H., Alam, M. Z., & Rahman, M. M. (2011). Host preference of red pumpkin beetle in a choice test under net case condition. *Bangladesh Journal of Zoology, 39*(2), 231-234.

York, A. (1992). Pests of cucurbit crops: marrow, pumpkin, squash, melon and cucumber. In *Vegetable crop pests* (pp. 139-161). London: Palgrave Macmillan UK.

Fruhwirth, G. O., & Hermetter, A. (2007). Seeds and oil of the Styrian oil pumpkin: Components and biological activities. *European Journal of Lipid Science and Technology, 109*(11), 1128-1140.

Halder, J., & Rai, A. B. (2020). Synthesis and development of pest management modules against major insect pests of pumpkin (Cucurbita moschata). *Indian Journal of Agricultural Sciences, 90*(9), 1673-1677.

Khan, M. M. H. (2012). Host preference of pumpkin beetle to cucurbits under field conditions. *Journal of the Asiatic Society of Bangladesh, Science, 38*(1), 75-82.

Everts, K. L. (2002). Reduced fungicide applications and host resistance for managing three diseases in pumpkin grown on a no-till cover crop. *Plant disease, 86*(10), 1134-1141.

Jaiswal, N., Singh, M., Dubey, R. S., Venkataramanappa, V., & Datta, D. (2013). Phytochemicals and antioxidative enzymes defence mechanism on occurrence of yellow vein mosaic disease of pumpkin (Cucurbita moschata). *3 Biotech, 3,* 287-295.

Singh, A. K., Mishra, K. K., Chattopadhyay, B., & Chakraborty, S. (2009). Biological and molecular characterization of a begomovirus associated with yellow mosaic vein mosaic disease of pumpkin from Northern India. *Virus Genes*, *39*, 359-370.

Mehl, H. L., & Epstein, L. (2007). Identification of Fusarium solani f. sp. cucurbitae race 1 and race 2 with PCR and production of disease-free pumpkin seeds. *Plant Disease*, *91*(10), 1288-1292.

Dhiman, A. K., Sharma, K. D., & Attri, S. (2009). Functional constitutents and processing of pumpkin: A review. *Journal of Food Science and Technology*, *46*(5), 411.

Kaur, S., Kaur, S., Srinivasan, R., Cheema, D. S., Lal, T., Ghai, T. R., & Chadha, M. L. (2010). Monitoring of major pests on cucumber, sweet pepper and tomato under net-house conditions in Punjab, India. *Pest Management in Horticultural Ecosystems*, *16*(2), 148-155.

York, A. (1992). Pests of cucurbit crops: marrow, pumpkin, squash, melon and cucumber. In *Vegetable crop pests* (pp. 139-161). London: Palgrave Macmillan UK.

Baiomy, F., & Tantawy, M. A. (2014). Comparison between the infestation rate of certain pests on cucumber and kidney bean and its relation with abiotic factors and anatomical characters. *Egyptian Academic Journal of Biological Sciences. A, Entomology*, *7*(2), 63-76.

Dimetry, N. Z., El-Laithy, A. Y., AbdEl-Salam, A. M. E., & El-Saiedy, A. E. (2013). Management of the major piercing sucking pests infesting cucumber under plastic house conditions. *Archives of Phytopathology and Plant Protection*, *46*(2), 158-171.

Zhang, S., Raza, W., Yang, X., Hu, J., Huang, Q., Xu, Y., ... & Shen, Q. (2008). Control of Fusarium wilt disease of cucumber plants with the application of a bioorganic fertilizer. *Biology and Fertility of Soils*, *44*, 1073-1080.

Pixia, D., & Xiangdong, W. (2013). Recognition of greenhouse cucumber disease based on image processing technology. *Open Journal of Applied Sciences*, *3*(01), 27-31.

Huang, X., Zhang, N., Yong, X., Yang, X., & Shen, Q. (2012). Biocontrol of Rhizoctonia solani damping-off disease in cucumber with Bacillus pumilus SQR-N43. *Microbiological Research*, *167*(3), 135-143.

Berdugo, C. A., Zito, R., Paulus, S., & Mahlein, A. K. (2014). Fusion of sensor data for the detection and differentiation of plant diseases in cucumber. *Plant pathology*, *63*(6), 1344-1356.

Roberts, D. P., Lohrke, S. M., Meyer, S. L., Buyer, J. S., Bowers, J. H., Baker, C. J., ... &

Chung, S. (2005). Biocontrol agents applied individually and in combination for suppression of soilborne diseases of cucumber. *Crop Protection, 24*(2), 141-155.

Alghali, A. M. (1991). Studies on cowpea farming practices in Nigeria, with emphasis on insect pest control. *International Journal of Pest Management, 37*(1), 71-74.

Marsh, R. E., Koehler, A. E., & Salmon, T. P. (1990). Exclusionary methods and materials to protect plants from pest mammals--a review. In *Proceedings of the Vertebrate Pest Conference* (Vol. 14, No. 14).

Jenkins, D. J., & Noad, B. (2003). *Guard animals for livestock protection: existing and potential use in Australia*. Orange, NSW, Australia: NSW Agriculture.

Belant, J. L., Seamans, T. W., & Dwyer, C. P. (1998). Cattle guards reduce white-tailed deer crossings through fence openings. *International Journal of Pest Management, 44*(4), 247-249.

Nuruzzaman, M. D., Rahman, M. M., Liu, Y., & Naidu, R. (2016). Nanoencapsulation, nano-guard for pesticides: a new window for safe application. *Journal of agricultural and food chemistry, 64*(7), 1447-1483.

Hendrikx, T., & Schnabl, B. (2019). Antimicrobial proteins: intestinal guards to protect against liver disease. *Journal of gastroenterology, 54*, 209-217.

Nasser, M. I., Zhu, S., Huang, H., Zhao, M., Wang, B., Ping, H., ... & Zhu, P. (2020). Macrophages: First guards in the prevention of cardiovascular diseases. *Life sciences, 250*, 117559.

Chaconas, G., Castellanos, M., & Verhey, T. B. (2020). Changing of the guard: How the Lyme disease spirochete subverts the host immune response. *Journal of Biological Chemistry, 295*(2), 301-313.

Lorang, J., Kidarsa, T., Bradford, C. S., Gilbert, B., Curtis, M., Tzeng, S. C., ... & Wolpert, T. J. (2012). Tricking the guard: exploiting plant defense for disease susceptibility. *Science, 338*(6107), 659-662.

Xiong, N., Xiong, J., Khare, G., Chen, C., Huang, J., Zhao, Y., ... & Wang, T. (2011). Edaravone guards dopamine neurons in a rotenone model for Parkinson's disease. *PLoS One, 6*(6), e20677.

## INSECT PESTS

### Mustard sawfly, *Athalia lugens proxima*

#### Damage symptoms

- » The larva feeds by nibbling at the edges of the leaf and eventually moves in towards the midrib.

- » As they gorge themselves, the grubs create multiple shot holes and sometimes even pierce the entire leaf.

- » They feed on the shoot's epidermis, which causes the plant to dry up and ultimately fail to produce seeds if the plant is mature enough.

#### Pest identification

- » Larvae are wrinkled and a bluish-green colour; they have eight pairs of pro-legs.

- » Larvae that are touched collapse to the ground in a fake death reaction.

- » The adult male has a black head and chest. The stomach area is orange. Clear, smoky, and dotted with black veins, the wings are translucent.

#### Management

- » The effectiveness of NSKE as both an antifeedant and a larvicidal spray was greatest at a concentration of 5%.

- » Sawflies were shown to be more susceptible to the larvicidal effects of Repellin (1%) and Neemark (1%) than to those of Margocide (0.1%) and Repellent (1%).

» Larval mortality was greatest when synthetic insecticides were used, specifically Triazophos + Decamethrin at 0.036% and Quinalphos at 0.05%.

» Carbaryl (Sevin 50 WP) sprayed at a concentration of 4 g/L in water, once per week.

## Aphids, *Brevicoryne brassicae, Lipaphis erysimi, Myzus persicae*

### Damage symptoms

» Crops at all stages of development and maturity are fair game.

» Aphids live in colonies, which can be observed crowded on the underside of leaves and on young stems.

» To get their nourishment, aphids feed on plant juices, plants with this disease curl and turn yellow.

» More damage is done to radish crops for seed than to those whose roots will be eaten.

» As the plants lose all of their vitality, their leaves and shoots curl and turn yellow before they eventually die.

### Favorable conditions

» The spread of their infection is aided by cloudy, humid weather.

### Management

» Use the allure of traps designed to ensnare.

» The best cost-effectiveness ratio was achieved with three 2-ml/L sprays of Monocrotophos or Endosulphan applied when there were 50 aphids (*Lipaphis erysimi*) on a plant.

» Pest control was achieved with the timely application of Phorate or Carbofuran to the soil at a rate of 1.5 kg active ingredient per hectare.

» When sprayed at a rate of one litre per thousand litres of water, Malathion has a residual impact that lasts for two to three weeks and is effective in killing a large percent of pests.

» One litre of nicotine sulphate in eight hundred litres of water is effective at roughly 21 °C.

### Root maggot, *Delia planipalpis*

### Damage symptoms

» The tiny fly larvae that burrow into radish roots to feed leave sticky brown tunnels behind.

» Some wilting of the foliage may occur before harvest.

### Pest identification

» White and tiny eggs.

» The larvae and nymphs of this species measure 8-9 mm in length, are yellowish to white in colour, and lack a distinct head and legs.

» Smaller than typical houseflies, an adult is 6-7 mm in length and has a uniform dark grey to brown colour.

### Survival

» Pupae, which resemble seeds in the soil, help them make it over the winter.

### Management

» Cover newly planted radish seeds with a floating fleece or net screen to prevent adult flies from depositing eggs on the foliage.

» Plants in the cabbage family can be rotated in to deter this pest. Planting non- mustard crops such as tomatoes, beans, peas, or even herbs in between your mustard plantings can also help deter this insect.

## DISEASES

### White rust, *Albugo candida*

### Symptoms

» The leaves and blossoms of infected plants are a prime target for the disease.

» Cotyledons, leaves, stems, and flowers may develop white pustules that eventually join together to form larger infected areas.

» The leaves may curl and grow thicker.

» Damaged leaves wither and fall to the ground.

» Deformed flowering shoots will only produce abnormal blossoms.

» On the underside of the leaves, a white powdery substance has formed in spots.

**Favorable conditions**

» The disease spreads rapidly in conditions of mild temperature and heavy humidity.

» It typically takes place when days are warm and nights are cool and rainy.

» Optimal growth temperature range is between 14 and 20 °C.

**Survival and spread**

» The pathogen survives the winter in two forms: mycelium in infected hosts and oospores in stag heads (galls developed on infected seed heads) or plant debris.

» Long lengths of time can pass without water for fungus.

» Contagious diseasees that can be dispersed by the elements or biting bugs.

**Management**

» Never plant infected seed.

» Injecting Captan or Thiram into the seed treatment process.

» Don't use sprinklers if at all possible.

» Rotate crops that are sensitive every three years.

» The inoculum can be reduced by destroying unhealthy crop debris.

» Keeping the area free of weeds and practising other forms of sanitation is also crucial.

» Disease can be avoided with the use of clean cultivation and disease-resistant plant kinds.

» Spraying with a Bordeaux mixture of 0.8% or Zineb 0.2% on a regular basis is an effective method of disease control.

**Black root, *Aphanomyces raphani***

**Symptoms**

» Infected radishes will have dark purple or even black skin surrounding the root.

» Lesions form, and the fungus spreads into the root's interior as the damage progresses.

» Rough, black-blue spots show up on the roots, and as they grow, they encircle the taproot.

» Roots are stunted where lesions occur.

» Roots are also affected by the black discolouration.

» Completely blackening radishes, on the inside and out, is a symptom of this sickness.

## Favorable conditions

» Growth is enhanced in damp, cool soil.

## Survival

» Fungi have a long period of survival in soil.

## Management

» Plant only sterile seedlings to avoid spreading disease.

» Add a lot of organic material to the bedding area to enhance drainage.

» Grow your plants where they will get plenty of air and water by using raised beds.

» Overwatering radishes might cause them to rot.

» Radish seeds should be planted in soils with a pH of 5.5 to 5.8.

» Non-brassica species should be rotated in every three to four years.

## Downy mildew, *Peronospora parasitica*

## Symptoms

» Fungal growth is seen on the underside of cotyledons and primary leaves when they are invaded in seed beds.

» A minor yellowing appears afterwards, just opposite the fungal development on the upper side of the leaf.

» Necrotic patches that start off as small, angular lesions on the upper leaf surface.

» Growth of white fluff on the leaf undersides

» When the cotyledon turns yellow, it falls off.

» Typically, older leaves remain and the infected patches get larger, turn brilliant yellow, and eventually turn tan and papery.

» The fungus causes damage to the plant's root system.

» The root is covered with large, dark, round patches.

» The final outcome is the total annihilation of plant life.

## Favorable conditions

» Viruses and bacteria thrive in conditions that are cool, moist, and humid.

## Spread

» Can be carried by the wind with relative ease.

## Management

» For at least two of every three years, rotate your crops such that you don't plant any cruciferous vegetables.

» Use clean seed beds separated from other crucifer producing regions and eradicate cruciferous weeds to maintain sanitary conditions.

» Make sure your plants get plenty of sunlight by selecting a location and arranging your plants in a grid.

» Killing off diseased plants.

» After harvesting, clean up completely.

» Maneb, Ridomil, or Aliette 0.2% Spray The disease should be checked on every 10 days for the first two weeks.

## Alternaria blight, *Alternaria raphani*

## Symptoms

» Little, round, slightly elevated lesions of varying shades of yellow, dark brown, and black first emerge on the leaves.

» Later on, petioles, stems, blooms, and fruit pods develop lesions.

» Sometimes the centre of lesions dries up and falls out, leaving a shot-hole pattern in the affected leaves.

» Necrotic patches grow in size as smaller ones merge into them.

» A loss of leaves could happen.

» There is a quick spread of infection during wet weather, and the entire pod may get infected, turning the styler end black and shrivelled.

» The seeds are infected by a fungus that grows into the pods.

» Faulty seed does not sprout.

**Favorable conditions**

> » Diseases are more likely to appear when temperatures are high and precipitation is high.

**Management**

> » Diseases are more likely to appear when temperatures are high and precipitation is high.

**Bacterial leaf spot,** *Xanthomonas campestris* **pv.** *raphani*

**Symptoms**

> » Any stage of development, from a seedling to a fully grown plant, is vulnerable.
>
> » Affected cotyledons on early plants have blackened borders and eventually die.
>
> » Spots on the leaves ranged from waterlogged to oily, and in some cases, the spots were surrounded by a thin yellow halo.
>
> » Leaks on leaves that are black and sunken and lengthy.
>
> » At the later stages of infection, water holes on the leaf margins become the entry point. Chlorosis causes the afflicted tissues to turn a yellowish colour.
>
> » Veins are discoloured a dark brown or black.
>
> » We drenched the soil with 10-12 kg of stable bleaching powder per hectare.

**Survival**

> » Transmitted via crop waste and contaminated seed.
>
> » It spreads by infected seeds.
>
> » Insects, rain, and so on can then spread the disease further after it has afflicted a plant.

**Management**

> » Grow only disease-free seeds.
>
> » Seeds are soaked in hot water (52 °C) for 30 minutes.
>
> » Seeds treated with various antibiotics at concentrations as low as 0.01%, such as Agromycin and Streptocycline.
>
> » Don't water the lawn from above.

» Soil-borne infections can be contained with just two years of crop rotation.

» Open up the space between the rows to improve ventilation.

» Application of a pine bark and compost mixture containing the biocontrol agent *Trichoderma hamatum* to the soil.

## References

Butani, D. K., & Juneja, S. S. (1984). Pests of radish in India and their control. *Pesticides*, *18*(5), 10-12.

Anooj, S. S., Raghavendra, K. V., Shashank, P. R., Nithya, C., Sardana, H. R., & Vaibhav, V. (2020). An emerging pest of radish, striped flea beetle Phyllotreta striolata (Fabricius), from Northern India: incidence, diagnosis and molecular analysis. *Phytoparasitica*, *48*, 743-753.

Nair, K. S. S., & McEwen, F. L. (1973). The seed maggot complex, Hylemya (Delia) platura and H.(Delia) liturata (Diptera: Anthomyiidae), as primary pests of radish. *The Canadian Entomologist*, *105*(3), 445-447.

Jaafar, N. A., Ahmed, A. S., & Al-Sandooq, D. L. (2020). Detection of active compounds in radish Raphanus Sativus L. and their various biological effects. *Plant Archives*, *20*(2), 1647-50.

Witkowska, E., Moorhouse, E. R., Jukes, A., Elliott, M. S., & Collier, R. H. (2018). Implementing Integrated Pest Management in commercial crops of radish (Raphanus sativus). *Crop Protection*, *114*, 148-154.

Duffus, J. E. (1960). Radish yellows, a disease of radish, sugar beet, and other crops. *Phytopathology*, *50*(5).

Tompkins, C. M. (1939). *A mosaic disease of radish in California*. US Government Printing Office.

Shin, T., Ahn, M., Kim, G. O., & Park, S. U. (2015). Biological activity of various radish species. *Oriental Pharmacy and Experimental Medicine*, *15*, 105-111.

Krause, M. S., De Ceuster, T. J. J., Tiquia, S. M., Michel Jr, F. C., Madden, L. V., & Hoitink, H. A. J. (2003). Isolation and characterization of rhizobacteria from composts that suppress the severity of bacterial leaf spot of radish. *Phytopathology*, *93*(10), 1292-1300.

# Tomato

## INSECT PESTS

### Fruit borer, *Helicoverpa armigera*

#### Damage symptoms

- » On the trifoliate leaves just below the uppermost flower cluster, they deposit tiny, single, spherical, whitish eggs.
- » When the eggs hatch, the first instar larvae feed on the leaves for a few days before moving on to the developing green fruits.
- » Afterwards, the larvae would burrow into the fruits, leaving their rear ends protruding.
- » The tomato fruit borer is a polyphagous pest that feeds on tomatoes, decreasing both productivity and price.
- » The cocoon is laid in the ground for the pupation process.

#### Pest identification

- » Eggs: Each egg is carefully crafted and is a creamy white colour.
- » The larvae are a range of colours from green to brown. Its body is covered in dark brown grey lines, and it also has white lines running laterally and a black band.
- » Brown pupae can be found in dirt, leaves, pods, and rotting crops.
- » The female moth is a pale brownish yellow and rather stocky adult, whereas the male moth is a pale greenish speck with a V shape. The centre of the forewing is a dark brown circle, while the rest of the wing is olive green to pale brown. The outermost margin of the pale smoky-white rear wing has a dark, almost black colour.

## Management

» Boring fruit is periodically harvested and destroyed by machines (3-4 times).

» Neem cake at a rate of 250 kg/ha for soil treatment.

» Both Neem soap and Pongamia soap, diluted to 1% and sprayed on the leaves of the plants at intervals of 10 days beginning at day 25 after planting.

» *Trichogramma brasiliensis, T. pretiosum,* and *T. chilonis* were released at a rate of 50,000/ ha over the course of five separate releases, the first of which occurred around the time flowering first began.

» Five 250 LE/ha applications of HaNPV beginning with the first spray during bloom initiation.

» The use of *Bacillus thuringiensis* (Dipel) sprays at a rate of 0.5 kg/ha, applied at 10- day intervals.

» Five applications of *Nomuraea rileyi* ($3.2 \times 10^8$ conidia/ml) and Triton X-100 (0.01%) were made on the ground once a week from dusk till dawn during the course of two months.

» Using marigold as a trap crop and applying sprays of Ha NPV @ 250 LE/ha at 28, 35, and 42 days after transplanting yields very effective management.

## Serpentine leaf miner, *Liriomyza trifolii*

## Damage symptoms

» White maggots begin leaf-mining immediately.

» For food, the tiny metallic insect punctures the leaf lamina.

» When the maggots eat their way through the leaf, they leave behind characteristic serpentine-shaped tunnels in the lamina.

» The leaves are withering and falling.

## Pest identification

» Little apodous maggots, orange and yellow in colour.

» Pupae: mines include yellowish-brown pupae.

» To an adult's eye, it seems a very light yellow.

## Management

» This problem often begins in the child's nursery. Thus, infected leaves should be removed either at the time of planting or no later than a week after transplanting.

» Neem cake, at a rate of 250 kg per hectare, should be broadcast over the planting furrows at the time of planting, and then again 25 days after planting, when flowers should be emerging.

» Use a 1% Neem soap spray or a 4% Neem seed powder extract spray at 15-20 DAP.

» Before spraying Triazophos 40 EC (1 ml) and Neem oil (7.5 g/L) on affected plants during times of high incidence, infected leaves should be removed.

» Avoid using insecticides carelessly so that natural predators might flourish.

» Adult flies can be caught using yellow sticky traps.

» Sheetal tomatoes were the least susceptible to pests, followed by Rupali, Rashmi, and Naveen.

## Whitefly, *Bemicia tabaci*

### Damage symptoms

» Both the nymphs and the adults eat on the underside of the leaves, distorting the new leaves in the process cause leaves to curl downward and wither

» Chlorotic leaves and withering are symptoms of a severe disease.

» Sooty mould is caused by honeydew, which is excreted by whiteflies.

» Tomato leaf curl virus is widely known to be spread by white fly.

### Pest identification

» Pear-shaped, pale-yellow, stalked egg.

» The nymph's scales are round and a bluish-green colour.

» A little white adult that resembles a scale.

### Management

» Eliminate the leaf curl plants from your garden.

» It is important to get rid of *Abutilon indicum* because it is a host plant for a different type of weed.

» Put out 12 yellow sticky traps per hectare to lure and kill pests.

» Employ hybrids that are immune to viruses.

» Use nylon nets or polyhouses to cultivate seedlings.

» Before planting, soak the roots in a solution of Imidacloprid (0.3 ml/L) or

Thiomethoxam (0.3 g/L) for 5 minutes.

» To prepare nursery plants for transplanting, soak protrays in the chemicals a day before planting.

» Root dipping systemic pesticides should be sprayed on the plants 15 days after planting.

» When signs of virus infection are spotted, remove the plants from the environment immediately.

## Thrips, *Thrips tabaci, Frankliniella occidentalis*

### Damage symptoms

» Damage by thrips can be identified by the wrinkling, cupping, and upward curling of the affected leaves.

» In later stages, the flowers will begin to fall.

» Marks of silver on the leaves.

» Flowers that have been cut too early.

» Dead flower buds.

» Tomato spotted wilt virus is spread by the insect.

### Pest identification

### Nymphs: Yellowish

» The mature form has a dark colour and has wing fringes.

### Management

» The infected trees and shrubs should be mechanically removed and burned.

» Try 15 traps per hectare with yellow sticky tape.

» Put out 10,000 *Chrysoperla carnea* larvae per hectare.

» Methyl demeton (25 EC) @ 1 L/ha, Dimethoate (30 EC) @ 1 L/ha, or Monocrotophos (2.0 ml/L) in water spray.

**Pinworm,** *Tuta absoluta*

### Crop losses

» Under greenhouse and field circumstances, it can impair fruit quality and cause a production loss of up to 90%.

### Damage symptoms

» Insects and animals that eat plants directly can cause damage to the plant health by munching on its leaves, stems, buds, calyces, young fruit, or ripe fruit.

» Extracting minerals from fruit stems, leaves, and seeds.

» Secondary infections can do further harm by entering damaged areas left by pests.

### Pest identification

» Eggs are tiny cylinders that range in colour from creamy white to a pale yellow and are around 0.35 mm in length.

» Larvae: White in colour when newly hatched, larvae of the second through fourth instars turn green or pale pink depending on the diet they've been provided (leaflet or ripe fruit, respectively).

» Moths reach maturity at 5–7 mm in length and 8–10 mm in wing spread, with silvery-grey scales, filiform antennae, light and dark segments, and well-developed recurved labial palps.

### Management

» Pick up and dispose of all infected plants and produce.

» Do not plant any Solanaceous plants after tomatoes.

» Transplant only vigorous seedlings.

» Maintain pheromone traps at the rate of 40 per hectare to capture and kill adult moths.

» Neem formulation (Azadirachtin 1% or 5%) @ 1.0 - 1.5 L/ha, Chlorantraniprole 18.5% SC @ 60 ml, Cyantraniprole 10% OD @ 60 ml, Flubendiamide 20% WG @ 60 ml, Indoxacarb 14.5% SC @ 100 ml.

## Aphid, *Aphis gossypii*

### Damage symptoms

- » Aphids have needlelike mouthparts that they use to pierce leaf veins, stems, growth tips, and flowers.
- » Because of this, flowers fall off and production decreases.
- » It causes new growth to be stifled and twisted.
- » Infested plants wilt and die from the top down.
- » There are a number of viral diseasees that aphids can transmit, and they have a tendency to jump from field to field.
- » They feed on the plant's sticky honeydew, which is full of nutrients.
- » Photosynthesis is stifled because honeydew attracts sooty mould fungus.
- » Tomatoes can be inhibited in growth and develop curling leaves if they are infested with aphids.

### Pest identification

- » Pear-shaped aphids (*Aphis gossypii*) can range in colour from yellow to dark green, but their cornicles (slender tailpipe-like appendages) are always a dark tint.

### Management

- » In the spring, aphids can be kept at bay with the use of a reflective mulch.
- » You might spray the area with insecticide soap or Neem oil to kill off the bugs.
- » Use a pesticide such as dimethoate, malathion, or imidachloprid.
- » The presence of natural predators like lacewings and ladybirds can help lower aphid populations.

## Red spider mites, *Tetranychus urticae*

### Damage symptoms

- » The tiny red mites cause the leaves to turn yellow and dry up because they feed by scraping the skin.
- » Leaves affected by this condition turn a rusty brown colour.
- » Silken webbing on the leaves indicates a severe larval infestation, which leads to leaf wilting and drying.

» The development of flowers and fruit is impeded.

» Mites are also a problem on fruit.

## Pest identification

» Hyaline, round eggs that are placed in clusters.

» The nymphs we see around here are sort of a lemony yellow.

» The mature specimens are a little, bright crimson in colour.

## Management

» Incidence tends to appear in clusters over a plot.

» The damaged leaves should be removed right away and sprayed with Dicofol 18.5 EC @ 2.5 ml or another acaricide, such as Wettable Sulfur 80 WP @ 3g/L.

» As mites are typically found on the undersides of leaves, here is where you should focus your spraying efforts.

» Instead of using chemical acaricides, try spraying a mixture of 1% Neem oil, 1% Neem soap, and 1% Pongamia soap.

» In extreme cases, you should begin by removing and disposing of any seriously affected leaves before combining an acaricide with a botanical (as described above) and spraying the area thoroughly.

» To avoid a widespread infestation of mites, it is best to practice integrated pest management (using marigold as a trap crop and spraying 4% NSKE).

# DISEASES

## Damping-off, *Pythium aphanidermatum*

## Symptoms

» There are two stages of damping-off in the nursery: before and after the tomato plants have emerged.

» Before to emerging from the earth, the seedlings are eradicated during the pre-emergence phase.

» All of the seedling's rot and die, including the young radicle and plumule.

» In the phase following emergence, the collar's immature, subterranean tissues are infected.

» The diseased tissue softens and absorbs a lot of water.

» Later, the seedlings' stems get shortened and they totter or collapse.

## Management

» Thirty days of solarization in the height of summer, accomplished using a polythene layer over the soil.

» When starting plants from seed, only use virus-free seeds.

» Seeds should be treated with 2 g/kg of Chlorothalanil, Captan, or Metalaxyl.

» Take bad apples and smash them.

» Before to the start of the monsoon, spray 0.2 % Metalaxyl or Mancozeb, and then use either 1.0 % Bordeaux mixture or 0.3 % Copper oxychloride every 8 to 10 days.

» Cymoxanil, Propineb mancozeb, and Chlorothalnil sprays, all at a concentration of 0.2%, are also useful.

» Fungus-culturing as a seed treatment with Trichoderma viride (4 g/kg of seed).

» Effective control of damping-off was achieved with seed treatment with *Trichoderma viride/Pseudomonas fluorescens/Azotobacter croococcum* and soil application of FYM enhanced with *T. viride*.

## Early blight, *Alternaria solani*

## Symptoms

» Little, dispersed, pale brown patches are the first visible symptom.

» Large and growing in concentric rings within necrotic tissue, they cause leaf blight by yellowing the leaves.

» There is also some girdling of the stem at the plant's base.

» Stem lesions are characterized by the extension and depression of a single side of a stem.

» Similar symptoms manifest on fruit, leading to fruit rot, stem end rot, or blossom end rot.

## Management

» To avoid spreading disease, plant from disease-free seed.

» Seeds should be treated with Thiram 75 WP at a rate of 2.5 g/kg.

» In order to reduce relative humidity in the plant canopy, a wider spacing of 90 x 60 cm is recommended.

» In order to prevent the spread of disease, it's important to remove lower leaves on a regular basis.

» Disease can be reduced through crop rotation.

» The blooming stage is the beginning point for spraying with either 0.2% Chlorothalanil, 0.2% Propineb, 0.2% Zineb, 0.3% Copper oxychloride, or 0.2% Iprodione + Carbendazim at 8-10 day intervals.

### Buckeye rot, *Phytophthora nicotianae* var. *parasitica*

### Crop losses

» In extreme cases, this disease can reduce fruit production by as much as 92%.

### Symptoms

» Fruits develop spots whose outermost rings are varying shades of brown.

» They might be tiny or cover a large area of the fruit's surface, resulting in fruit rot and a significantly reduced harvest.

» Overnight, a single contaminated fruit can cause the entire basket to decay.

### Management

» Use the ridge and furrow planting method.

» Once your plants have reached the flowering stage, stake them.

» In the beginning of monsoon season, cut all of the lower leaves to the ground (within 15–20 cm).

» Get rid of the spoiled fruit.

» Bordeaux spray (1%), Propineb spray (0.2%), Dithionon spray (0.2%), or Metalaxyl

» Mancozeb spray (0.2%).

### Fusarium wilt, *Fusarium oxysporum* f. sp. *lycopersici*

### Symptoms

» Veinlet clearing and chlorosis in the leaves are the earliest signs of the disease.

» During the period of a few days, the plant could wilt and die from the loss of its younger leaves.

- » The petiole and leaves will soon droop and wither.
- » As young plants show symptoms, they rid themselves of veinlets and lose their petioles.
- » In the wild, afflicted leaflets wilt and die when the lower leaves turn yellow.
- » This pattern of symptoms persists in succeeding leaf generations.
- » Discoloration of the plant's stem tissue is common.
- » The vascular system begins to brown at a later stage.
- » he growth of plants is hindered, and eventually they perish.

## Management

- » Thiram (2.5 kg/ha) should be used to treat seeds.
- » Plants at the nursery should be checked for wilt on a regular basis.
- » The diseased vegetation must be uprooted and disposed of.
- » Carbendazim (0.1%) or Benomyl (0.2%) should be sprayed on the soil before planting.
- » Planting cereals or other non-host crops every few years can help minimise the disease inoculum.

## Septoria leaf spot, *Septoria lycopersici*

## Symptoms

- » Any point in the plant's development is vulnerable to attack.
- » Little, grey, water-soaked, circular leaf spots with a dark border are a hallmark of the disease, and they tend to cluster on the undersides of lower leaves.
- » When blown up, their light grey centres and dark brown edges contrast dramatically.
- » There is a risk of yield losses and sun blistering of fruits if similar spots form on the flower calyx and stem (rarely on the fruits).

## Management

- » Damaged plant tissue is cut off and thrown away.
- » Thiram 75 WP, at a rate of 2 grammes per kg of seed, can be used to treat seeds.
- » A 0.2% spray solution of Chlorothalanil, Propineb, Mancozeb M, or Iprodione + Carbendazim should be used every 10-14 days.

» The disease can also be effectively controlled by spraying either 0.4% Copper oxychloride, 0.2% Mancozeb, or 0.1% Carbendazim.

## Bacterial wilt, *Ralstonia solanacearum*

## Symptoms

» Symptomatic of bacterial wilt is the sudden and total wilting of otherwise healthy, mature plants, with no prior yellowing of the leaves or collapse of the entire plant.

» Before the plant completely wilts, the lower leaves may fall.

» Infection is mainly restricted to the vascular system.

» Tissues may get discoloured and turn a yellowish-brown colour if the condition worsens to the point where it invades the cortex and the pith.

» There is a white track of bacterial slime that can be observed oozing from cut plant components when they are submerged in clear water.

» It causes rapid wilting and death of the plant, leading to complete destruction.

## Survival and spread

» Irrigation water, contaminated soil, infected seeds, plant debris, contaminated farming instruments, contaminated footwear, and infected transplants all play a role in the spread of this disease.

## Management

» It is recommended to apply 15 kg/ha of bleaching powder and groundnut cake to the soil before planting.

» Cowpea-maize-cabbage, okra-maize, maize-maize, and finger millet-brinjal (Pusa Purple Cluster)-French bean are all examples of crop rotations.

» omatoes were grown with sorghum, corn, onions, garlic, and rigolds.

» To improve tomato yields, we use brinjal rootstocks like Pusa Purple Cluster and Dingra Multiple Purple to create grafts for tomato plants.

» Streptocycline (1 g/40 litres of water) applied to seedlings and left on them for 30 minutes provides protection throughout the critical first few days of growth.

» Seed inoculation with 1010 colony forming units (cfu) of *Pseudomonas fluorescens, P. aeruginosa,* and *Bacillus subtilis.*

» Inoculating tomato plants with *Trichoderma viride* and *P. fluorescens* beforehand helps to prevent root rot.

» Plant seedlings are dipped in a non-pathogenic *Ralstonia solanacearum* (Av 10) strain to prevent root rot before being transplanted.

» *P. fluorescens/B. subtilis* root dip for seedlings.

» Planting *Glomus mosseae* in the ground.

» All of the following varieties are immune to bacterial wilt: Arka Abha, Arka Alok, Arka Shreshta, Arka Abhijit, Megha, Shakthi, Sonali, Sun 7610, Sun 7611, BT-1, BT-10, and LE 79-5.

» *Pseudomonas fluorescens* (1010 cfu/ml) seed treatment and the integration of sunnhemp (green manure crop) into the growing system significantly increased fruit yield from 7.8 to 25.5 MT/ha and decreased bacterial wilt incidence from 55.6% to 6.75 % from the high wilt incidence and low yield seen in the control group.

## Bacterial leaf spot, *Xanthomonas campestris pv. vesicatoria*

### Symptoms

» Little, brown, water-soaked, circular dark spots are seen on the leaves of infected plants, encircled by a yellowish halo.

» Defoliation is a severe risk for older plants infected with leaflets since they primarily affect older leaves.

» Most obviously, this disease manifests itself on the green parts of the fruit. It starts with tiny, water-soaked spots that gradually become elevated and grow to be between a sixteenth and a quarter of an inch across.

» These lesions' centres develop wavy and scabby with a pale brown, sunken appearance.

» Fruits that are fully mature are immune to the disease. Bacteria infect the seed's surface and stick there for a while.

» It can also live on volunteer tomato plants and plant debris that has been infected.

### Management

» It is imperative to constantly use disease-free seed and seedlings.

» In order to grow successfully, stick to clean methods.

» So that you don't have to deal with the leftovers from last year's harvest, it's best to rotate your crops with ones that aren't hosts.

» The disease can be kept under control by spraying Agrimycin-100 (100 ppm) three times, once every 10 days.

» Use a 1% Bordeaux mixture with three sprays.

» Streptocycline 150 ppm should be sprayed once.

» Copper oxychloride 0.3 % spray.

» Plants with higher tolerance, such LE-5, should be used.

**Mosaic disease, *Tomato Mosaic Virus* (ToMV)**

**Symptoms**

» When plants are first infected, they commonly show symptoms including light and dark green mottling on the leaves and the wilting of new leaves during bright days.

» Symptoms of a disease are leaflets that are smaller, puckered, and distorted from their typical shape.

» "Fern leaf" symptoms occur when the leaflets get depressed.

» The diseased plant looks stunted, pallid, and frail.

» The fungus can infect both plants and food.

**Spread**

» Clothing, hands, tools, and tools used on infected plants can all spread the infection to healthy plants.

**Management**

» It's important to choose disease-free plant seeds while planting a garden.

» It's important to give the seeds a good wash and let them dry in the shade.

» For the previous 48 hours prior to planting, soak the seeds in a Trisodium phosphate solution of 0.3%.

» Before planting, soak the seedlings for 10 minutes in a 0.2% Trisodium phosphate or Monocrotophos solution.

» Any diseased seedlings or mature plants must be carefully culled or destroyed.

» Seedlings infected with tobacco mosaic virus should not be used for transplantation.

» Non-solanaceous crops should be rotated in with the solanaceous ones.

# Leaf curl disease, *Tomato Leaf Curl Virus* (TLCV)

## Symptoms

» Plants affected by leaf curl disease are severely stunted and their leaves become chlorotic, rolled under, crinkled, curled, twisted, and chlorophyll-less.

» Curling and a mild yellowing of the leaves are other characteristics seen in newly emerged leaves.

» The older the leaves, the more leathery and brittle they become

» Virus infection causes a variety of symptoms in plants, including stunting, internode shortening, a bushier look, and a lack of flowering and fruiting.

» Infected plants become whitish and generate an abundance of side branches, giving them a bushy appearance.

» Infected plants are either either infertile or completely sterile.

» When plants are infected late, they still produce fruit, but the fruit is little, misshapen, and unsellable.

## Spread

» Whiteflies are responsible for the spread of this disease (*Bemisia tabaci*).

» Try cultivating a crop like Avinash-2 (T), Hisar Anmol (H-24), Hisar Gourav, JK Asha, Nandi, Abhinav, Vaibhav, Sankranti, or Mruthyunjaya-3, which are all resistant to or tolerant of leaf curl.

## Management

» The diseased vegetation must be uprooted and disposed. Weeding and earthing up operations can help reduce the spread of this disease by removing alternate or collateral hosts that may be harbouring the virus that causes it.

» Using a bioagent (*Encarsia, Trialeurodes vaporarium*) to control white flies helps lessen the spread of disease.

» Seeds are treated for 25 minutes in either hot water kept at 50°C or a 2% Trisodium phosphate solution.

» It is recommended that a nylon net be used to protect nurseries from whiteflies, the vectors of leaf curl.

» Plants infected by leaf curl must be uprooted and destroyed.

» Planting maize, sorghum, and bajra as borders or barriers around your tomato plot 50-60 days before you plan to transplant the tomatoes can deter viruliferous whiteflies.

» Sticky yellow traps are an option.

» Use 0.1% foliar sprays of either Monocrotophos or Dimethoate every 10 days.

**Spotted wilt disease,** *Tomato spotted wilt virus* **(TSWV)**

**Symptoms**

» The most recognisable signs are the rapid browning of leaves and the subsequent halt in plant growth.

» Later on, distortions appear on the leaves and necrosis sets in.

» The leaves have a copper color. The damaged portions often become necrotic and the bronzed parts roll inward.

» Infected plants' fruits are covered in numerous circular spots about 1.2 cm in diameter.

» Bands of red and yellow appear alternately on ripe fruits, and can be used as a diagnostic indicator of the presence of viruses.

**Spread**

» Triffids are responsible for transmitting the spotted wilt virus (*Thrips tabaci, Frankliniella schultzi* and *F. occidentalis*).

**Management**

» Grow seedlings in a greenhouse with a nylon net.

» As soon as you see signs of disease, pull up and burn the damaged plants.

» Weeding and earthing up operations can help reduce the spread of this disease by removing alternate or collateral hosts that may be harbouring the virus that causes it.

» Spraying 0.15% Monocrotophos/10% Adathoda leaf extract helps prevent the spread of disease.

» Foliar applications of 4% NSKE, 0.1% Monocrotophos, or Dimethoate every 10 days are effective.

» Mulching the soil with a reflective aluminum-coated plastic mulch greatly decreased disease incidence by discouraging the thrips vectors that spread the disease.

# References

Lange, W. H., & Bronson, L. (1981). Insect pests of tomatoes. *Annual review of entomology*, *26*(1), 345-371.

Gajanana, T. M., Moorthy, P. N., Anupama, H. L., Raghunatha, R., & Kumar, G. T. (2006). Integrated pest and disease management in tomato: an economic analysis. *Agricultural economics research review*, *19*(2), 269-280.

Baicu, T. (1996). Principles of integrated pest and disease management. *Principles of integrated pest and disease management*.

Flint, M. L., & Van den Bosch, R. (2012). *Introduction to integrated pest management*. Springer Science & Business Media.

Kennedy, G. G. (2003). Tomato, pests, parasitoids, and predators: tritrophic interactions involving the genus Lycopersicon. *Annual review of entomology*, *48*(1), 51-72.

Sharma, D., Maqbool, A., Ahmad, H., Srivastava, K., Kumar, M., Vir, V., & Jamwal, S. (2013). Effect of meteorological factors on the population dynamics of insect pests of tomato. *Vegetable science*, *40*(1), 90-92.

Chavan, R. D., Yeotikar, S. G., Gaikwad, B. B., & Dongarjal, R. P. (2015). Management of major pests of tomato with biopesticides. *Journal of Entomological Research*, *39*(3), 213-217.

Hanssen, I. M., Lapidot, M., & Thomma, B. P. (2010). Emerging viral diseases of tomato crops. *Molecular plant-microbe interactions*, *23*(5), 539-548.

Vasudeva, R. S., & Sam Raj, J. (1948). A leaf-curl disease of tomato. *Phytopathology*, *38*(5).

Ong, S. N., Taheri, S., Othman, R. Y., & Teo, C. H. (2020). Viral disease of tomato crops (Solanum lycopesicum L.): an overview. *Journal of Plant Diseases and Protection*, *127*, 725-739.

Sanoubar, R., & Barbanti, L. (2017). Fungal diseases on tomato plant under greenhouse condition. *European Journal of Biological Research*, *7*(4), 299-308.

Shenge, K. C., Mabagala, R. B., & Mortensen, C. N. (2010). Current status of bacterial-speck and-spot diseases of tomato in three tomato-growing regions of Tanzania. *Journal of Agricultural Extension and Rural Development*, *2*(5), 84-88.

Printed in the United States
by Baker & Taylor Publisher Services